# SHAPES OF IMAGINATION

# SHAPES OF IMAGINATION

Calculating in Coleridge's Magical Realm

George Stiny

The MIT Press
Cambridge, Massachusetts
London, England

The MIT Press would like to thank the anonymous peer reviewers who provided comments on drafts of this book. The generous work of academic experts is essential for establishing the authority and quality of our publications. We acknowledge with gratitude the contributions of these otherwise uncredited readers.

This book was set in ITC Stone Serif Std and ITC Stone Sans Std by New Best-set Typesetters Ltd. Printed and bound in the United States of America.

Library of Congress Cataloging-in-Publication Data

Names: Stiny, George, author.
Title: Shapes of imagination : calculating in Coleridge's magical realm / George Stiny.
Description: Cambridge, Massachusetts : The MIT Press, [2022] | Includes bibliographical references and index.
Identifiers: LCCN 2021046826 | ISBN 9780262544139 (paperback)
Subjects: LCSH: Shapes—Design—Art. | Geometrical models—Design—Art. | Creation (Literary, artistic, etc.)—Mathematics. | Aesthetics—Data processing.
Classification: LCC QA445 .S765 2022 | DDC 516—dc23/eng/20211214
LC record available at https://lccn.loc.gov/2021046826

10  9  8  7  6  5  4  3  2  1

To Lionel March—
in pursuit of Alberti's delight in lines and numbers.

## On the imagination, or esemplastic power

The IMAGINATION then I consider either as primary, or secondary. The primary IMAGINATION I hold to be the living Power and prime Agent of all human Perception, and as a repetition in the finite mind of the eternal act of creation in the infinite I AM. The secondary I consider as an echo of the former, co-existing with the conscious will, yet still as identical with the primary in the *kind* of its agency, and differing only in *degree*, and in the *mode* of its operation. It dissolves, diffuses, dissipates, in order to re-create; or where this process is rendered impossible, yet still at all events it struggles to idealize and to unify. It is essentially *vital*, even as all objects (*as* objects) are essentially fixed and dead.

FANCY, on the contrary, has no other counters to play with, but fixities and definites. The Fancy is indeed no other than a mode of Memory emancipated from the order of time and space; and blended with, and modified by that empirical phenomenon of the will, which we express by the word CHOICE. But equally with the ordinary memory it must receive all its materials ready made from the law of association.

—Samuel Taylor Coleridge

# Contents

# Preface

This book consists of one longish essay, "Shape Grammars: Seven Questions and their Short Answers," and three supporting exhibits that are each self-contained yet invariably drawn into one another and into my seven answers—"Theory," "Observations," and "Pedagogy." In my questions and answers, I make the case for visual calculating with rules in shape grammars, and art and design, as a single enterprise in which everything aligns coequally. In art, there's the complete sweep of visual things, and in design, a steady reminder of making and material, and function and use. The agreed approach to calculating, and art and design is to run the former through the latter, so that art and design conform fully to what calculating allows in terms of given axioms and prior invariants. The claim today is that calculating in computers includes everything we see and do, according to custom and plan. In fact, computers "take away the looking" to open up art and design to art experts and everyone else, and to barter and trade what the fickle eye feels, for steady truth. (This is how the Rijksmuseum explains the use of computers in its current imaging and restoration of *The Night Watch*. The popular title of Rembrandt's painting may be mistaken in an unexpected way. As soon as computers "take away the looking," there's nothing to watch nor anything to see—night or day.) The hoo-ha around this is incredible—that the extent of known to new is within the reach of data and machine learning, that there are "training sets" to define it, and that nothing else is needed. Time will tell how far this vast ambition goes without slowing down. My approach in shape grammars is just the opposite—seeing informs calculating in art and design, so that calculating overtakes computers and what they hold, whether in logic or with data and learning. This is the only way I know to make calculating, and art and design coincide. There are many strands to this that bring in strange and wonderful things. These strands may seem very different, maybe irreconcilable—seeing that they're really not and how they overlap and interlace points to my key aims and goals. It will soon be apparent that I find scant to recommend in proven disciplines—even sporting different hats on different days, a physicist now and a painter tomorrow. Nor am I drawn to interdisciplinary work that only serves to reinforce separate methods and results in a combinatorial scheme, where everyone is an expert in his/her own area with no incentive to erase known boundaries, so that new configurations emerge and re-form freely in a totally seamless whole. This draws in diverse fields of interest to put them together in a single synthesis (sketch) that's inherently ambiguous—for re-division and ongoing description outside of familiar taxonomies (trees, topologies, and graphs, etc.) that rely on fixed differences to map underlying relationships. (The kind of relationship that strikes me most is conspicuous in metaphor and figuration, and runs throughout drawing and painting, and visual expression in general. More than a hundred years ago, the analytic philosopher G. E. Moore described it rather neatly, in an understandably wary account of something else—"organic unity" for living/purposeful things, that eludes "mechanical" relations like the combinatorial ones favored in logic and indispensable for computers and taxonomies. Somehow, reuse produces a positive result, as it dodges the egregious charge of plagiarism. Moore's description begs to be tried anew, to skew its original drift and purpose, and to steer past its provenance. Each relationship of the kind I have in mind "is supposed to be a relation which alters the things it relates, so that it is not they, but some [number of] other things which are related." Maybe this is an example of itself that shows what it means, as Moore's intentions evolve into my meaning to change the one into the other. It's funny how inconstant words can be, even the philosopher's. Or maybe it's Oscar Wilde, who extends meaning beyond intentions in his aesthetic formula for beautiful form, to pictures and poems, etc.—to see things as in themselves they really are not. This is limitless for most things, and especially in pictures; simple shapes are enough, drawing them with merely a handful of straight lines—surprises burst into view, far too numerous to overlook and too fascinating to ignore. A pair of polygonal L's

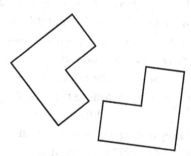

in another relation

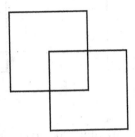

is two squares or three, or the center square in a symmetrical octagon with twin con-
cavities, staggered on the left and right. There's plenty of anxiety not knowing what's
next, and plenty of pleasure and delight, too, trying to find out—I'm keen on sliding
the L's into one another on their common diagonal to see what kinds of shapes I'm
able to get. I can fix the L's and move them back and forth on this axis, so that the
$\sqrt{2}$-distance x measured between their inflected vertices varies without any breaks, to
describe new relations parametrically. It's the same with separate puzzle pieces or better, overlaying transparent tiles or layers. When the two L's interact at −1, 0, and 1, they
look like this

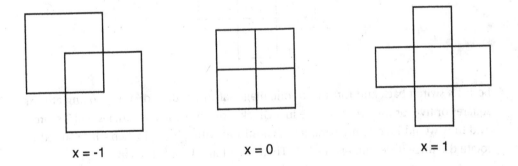

It helps to fill in the gaps on both sides of 0, to see how different things can be. Try
intercalating at −1/2 and 1/2—the L's are independent/discrete with non-overlapping
edges, in the way it is everywhere except at 0, even as their edges intersect copiously
in distinct arrays, four times on a virtual straight line, and six times on the sides of the
large, center square. The L's are quickly resolved when I use my finger to trace these
intersections in sequence—they just pop out of what, at first, may seem confusing

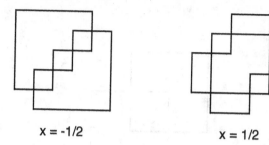

x = -1/2    x = 1/2

Four additional places complete my ninefold survey, at 3/2 and 2, and anywhere less than −1 or greater than 2. Of course, L's aren't everything—if I let them change freely, in the way I did to start, other things unfold in the wanton sweep of my eye. It's amazing all there is to see in unrestricted free play that's casually censored/lost, as parametric definitions fix things/parts for good in descriptions/representations that allow for position, angle, size, and other properties like color and material to vary. The agile eye isn't limited in this way; it steadfastly seeks for more—in this relation

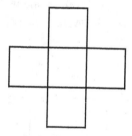

both L's switch back and forth from one diagonal to another, or am I looking at four squares or five, or the center square in a Greek cross, or varied rectangles and T's, etc.? And how would I recast my parametric definition with L's—and when—if I wanted to rotate the three little squares at −1/2? There aren't any L's in the shape

What happens to the definition if I decide to vary the size the large square in the two
L's at 1/2, dilating the square and contracting it in the series

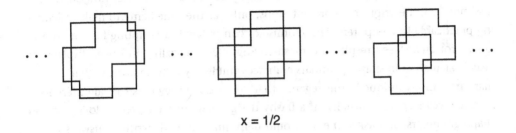

$$x = 1/2$$

and then try to rotate the four/two smaller squares? The instability in this highlights the
ambiguity that's everywhere in shapes. It's uncanny when definite things go together
in one way, only to change into different things arranged in another way. The magic
is hard to predict—I'm free to see, in an open-ended process in which nothing prior is
given or known. This is a strong start on what rules in shape grammars do when they're
used for visual calculating, to bring in the vast ambiguity that's at the quick of art and
design. There's simply no telling what my kind of relationship as in itself it really is.
Logical coherence and scientific rigor, however, are paramount—so there's little concern that this might swerve wildly out of control. The inconstant eye with its unsettled
results is easy to pooh-pooh for proper relations in computers that are defined for
constant/fixed/invariant things. These days, visual expression, etc. are ignored in computers, including object-oriented systems and machine learning, and when things stay
the same as they merge in sets and lists, and other data structures. This welcomes parametric definitions and equally, John von Neumann's "visual analogies" and my "spatial
relations"—in "Seven Questions," the three entwine to show why computers as they're
normally used aren't shape grammars. Surely, it's more productive to be a scientist than
an artist, or at least to try and be coherent and rigorous most of the time—some hats
are useless in pursuit of relationships that hew to the proven norms in logic and science.) I'm happy to go from math and logic to painting and poetry in the same breath,
in a single sentence and thought—the gaps dividing fields fade away leaving barely
a trace. It presumes far too much to ask you to understand how any of this works in
everyday practice, before I've posed my seven questions and you've had a chance to
read my seven answers, where everything changes into something new as it combines.
Isn't that the real point of a preface, as an advertisement to join an adventure—to get
you to go on reading? Shape grammars are my way of showing how rules work in order

to calculate visually, in broad detail with a kind of algorithmic certainty. Shape grammars do the trick without giving up any of the ambiguity inherent in whatever you and I see, in a fleeting glance or that's patiently observed—no matter how slight this loss may be in descriptions/representations, only for the time being to make calculating practicable in computers. The promise of computers is a tempting lure in art and design for guaranteed results and mesmerizing effects, dazzling in novel displays, as visual calculating in shape grammars embraces ambiguity fully. Evidently, shape grammars are entirely unique in this regard—that's the reason they make a real difference.

A few years ago, a journalist at a flashy design magazine wanted to do a piece on shape grammars, and asked me if I would help him out, with written answers to the seven questions in my essay. I agreed, but nothing seemed to result from the effort—I never saw my original, very short answers, or anything like them in print. The seven questions, however, stuck—they seemed to be a neat way to highlight some of what I'd been thinking about since the publication of my previous book, *Shape: Talking about Seeing and Doing* (Cambridge, MA: MIT Press, 2006). I haven't altered the questions in any substantive way, or the order in which they were posed, although my answers have grown luxuriantly. In writing them out, I've tried to make everything into an organic whole. This is evident in my enthusiasm for Samuel Taylor Coleridge, that's complete in one of my major themes—that shape grammars "in the *mode* of [their] operation" assimilate "imagination" and "fancy" in the way Coleridge intends them. (To the Romantics, imagination is the root of feeling and poetry; Coleridge's kind relies on "esemplastic power." Fancy keeps to "mechanico-corpuscular" invention; today, "combinatoric play.") In order to meld this theme and sundry other ones, so that in sum they're coadunate, I've used paragraphs sparingly, with long and short parentheses inserted throughout. My earlier parentheses for the ambiguous (organic) relationships I like is a good example—at the finish you look for the start, reading in reverse to see anew. It all blends naturally, for the artist and critic in close proximity and the artist and logician at remote poles. The sweep is vast and extravagant, and at times, may seem crazy. For example, the jump/swing from fancy to open-ended imagination in Coleridge entails "negative capability" in John Keats, and shows why "fact & reason" never tame surprise and doubt in art and design—negative capability is the only recourse to unchecked ambiguity, and the anxiety of looking when nothing is sure. Imagination without negative capability is intolerable, and negative capability without imagination is silly. (Keats doesn't count Coleridge among the great poets—negative capability eludes him, "being incapable of remaining content with half-knowledge . . . with a great poet the sense of Beauty overcomes every other consideration, or rather obliterates all consideration." This may be so, but negative capability isn't for everyone. It's hard to give up on definite answers in an equivocal world—"modern judgment, it seems, can accept either Keats or Coleridge—not both." Maybe modern judgment is why computers distinguish symbols, at least 0's and 1's. Or maybe it's the reverse—real differences stick, reliably with computers, and data and facts to inform judgment, so that optimization replaces whim and personal choice. Whichever way this goes, half-knowledge isn't an acceptable result—modern judgment and computers deter negative capability and with it, imagination. I guess it's best to keep Keats and Coleridge apart, mapped to 0 and 1—anything else is too dodgy, only for shape grammars that revel in ambiguity, and imagination with its esemplastic power.) And in the same vein, Wilde's aesthetic formula and the claims of the "critical spirit" in beautiful form, in varied modes of reverie and mood, mark the path von Neumann charts through the vagaries of personality, to probe the limits of calculating in pictures and the Rorschach test. Diverse ideas, and themes of all sorts commingle mutually and grow as one, without individuation and segmentation—that's left for reading in the way shapes have parts that are left for seeing. In this process, things rarely stay the same. There's no overarching outline, underlying structure, or plan or hierarchy to arrange and order what I say; it's mostly talking freely about what I see, and that this alters at will, as I go on. Moreover, there are some adventurous notes that overlap, tangled even in straightforward enumeration. This may be the place to start, to see how outwardly separate ideas combine organically in shape grammars. I've tried these rhetorical methods on a wide range of readers—parametrically, for those familiar with shape grammars and not, and for those at different points in between these extremes—and they seem to work effectively for nearly everyone. Each of the following three exhibits has its own pattern of growth, too. In the first exhibit, I do the math for shapes and shape grammars—I've always felt a strong moral obligation for this, if only to prove that visual calculating in shape grammars works just as rigorously as any other kind of calculating that's been set out entirely, in convincing formal detail. Shapes aren't numbers or symbols, but this doesn't mean that calculating with shapes isn't calculating equally the same—and it isn't necessary to describe or represent shapes in numbers and symbols for this to work. It's odd that anyone would think otherwise—seeing is hard to argue with—yet somehow, seeing that the math can be written down offers a guarantee that makes the case irrefutable to everyone. (It's telling how readily the pieces fall into place in a key analogy implicit in "Seven Questions." The analogy links Coleridge's ratio that compares imagination and fancy, and my ratio for shapes and symbols. These pairs correspond in the following way—

imagination : fancy :: shapes : symbols

The analogy is a productive template for other analogies, too—loads of them. I highlight new pairs twice outside of this parentheses, in my preface and acknowledgements, but the total score is dozens with many left to find. Of course, not all pairs work; some are similarities—artist and critic, Keats and Coleridge, and von Neumann and Wilde. Shape grammars capture the reciprocal relationship in every pair that aligns with shapes and symbols—maybe artist and logician—in terms of the dimension i, where i is 0 for points, and with extension in space and boundaries, 1 for lines, 2 for planes, 3 for solids, etc. The dimension i is 0 for symbols, because they behave independently like points when they combine, and greater than or equal to 0 for shapes, although some distinctions improve when i isn't 0. The analogy unfolds neatly in the usually dubious ratio of 1 to 0, that is to say, for lines and points.) In the second exhibit, I show why seeing and shapes are invariably unstable, and why the inherent ambiguity makes such a huge difference in shape grammars. My presentation relies first and foremost on line drawings, and how they change in funny ways, as I try out a varied mix of rules for parts, transformations, and boundaries in a recursive process—by looking repeatedly and re-drawing. Still, looking tends to be enough. The asymmetry is unavoidable—the eye is the highest judge in visual calculating. And finally in the third exhibit, I put everything together to show how the first two exhibits add up for teaching in art and design, not in a typical combinatorial/set-theoretic (mechanical) sum in which things remain independent, each with an invariant identity, but in a visual (organic) sum in which things disappear as they meld and alter. My trio of exhibits traces another route to traverse my seven questions, among surprisingly many alternate ways that seem worth taking, elucidating and scenic. It strikes me that all roads meander pleasurably through green fields in a lush and overgrown landscape to a common destination—yes, shape grammars. And why not? Shape grammars are an inclusive way of calculating that takes in how the mind thinks in terms of descriptions and representations, and amazingly, the myriad myriads of extravagant things the eye sees that elude prior analysis—or simply borrowing from Keats, what imagination holds for artist and poet to "see, where Learning hath no light." (There's little doubt that learning has greater reach today than 200 years ago, especially with statistics, data, and computers, but serious omissions persist—descriptions and representations are incomplete in computers and not, and this isn't something that can be completely debugged or finally filled in.) No one seems to care very much about the kind of seeing I have in mind, and what it does in visual calculating. Seeing is taken mainly for granted, despite the fact that there's nowhere for genuine thinking to go without it. (This may boost the claim that knowledge rests on imagination—for the Romantics and signally, Coleridge.) Everything follows from perception, not the other way around from reasoning and thought, even when the latter influences the former. The conditions for seeing to kick in seem beyond knowing; there's no clear path to the magical (aesthetic/creative) *that*, that happens automatically before thinking takes off and whenever it spins and turns to indulge a sudden intuition that promises a new purpose and fresh goals. Seeing works freely in shape grammars because they supersede thinking, tying imagination and fancy in analogy to shapes and symbols. And this extends to the rote results of learning and the constancies in computer/mechanical relations, to analysis, data, description, representation, and the vast number of things that come with some kind of fixed/unalterable structure—tellingly, visual analogies and spatial relations. The proof plays out in my seven questions and their short answers, and with no detail left out, in *Shape*.

I take seeing and calculating seriously to show how the one (seeing) expands the other, so that together they include art and design—and to show why seeing is key in calculating for rules to add anything of value and aesthetic import to creative activity. I'm often asked why I bother to tie in such different things—they're fine apart, the way they're meant to be. Isn't the mathematics of shape grammars enough? It's largely unfinished with plenty to do—come on, you should continue with that. The math alone merits your full attention and effort—why worry about art and design? My answer is twofold. First, I've done a lot of math already, and try to encourage others to carry on, in this remarkably enjoyable pursuit. And there's a growing number who can, well versed in unsolved problems, and ready to allow in what seeing in art and design implies to find effective solutions. Second, and of greater importance, shape grammars from the start have been part and parcel of a full-fledged aesthetic enterprise in open-ended visual perception, and other modes of sense experience. This relationship is something to affirm overtly once and for all, here via Coleridge, and also in league with von Neumann and Wilde, among many others inclined in similar ways. Then, everyone in art and design can take a deep breath and finally relax, join in and not worry about the math—and that art and design are at root calculating with rules. The math is just another way of talking that helps explain in marvelous detail what you and I see and do. It isn't only numbers and measurement, kitschy origami, and flashy pictures for messy data, difficult equations, and impossible Julia sets. Pictures may be useful in science but invariably, this is a recipe for bad art, betrayed in Wilde's aesthetic formula that's indifferent to intentions. Useful things are ordered, classified and sorted uniquely by use—defined, described, represented, or simply named. Useless things reverse this; in untold ways, they are as in themselves they really are not—"All art is quite useless." The idea is to link visual calculating in shape grammars, and art and design reciprocally, so that nothing can take away the looking—neither use nor any other prior description/representation. Useless things are infused with life once seeing (perception) is added to calculating; it's "poetry in motion" as they alter, this to that to this, etc.—intercalating at will, with new parts and pieces, that may overlap and not. This raises another question about computers and aesthetics. Recently, there's been a rash of studies (mostly in PhD dissertations) on the history of computers and what they do in/for art and design, in particular, in architectural practice and landscape design. These all end pretty much in the same way—maybe they have important uses, but computers fail to reach the visual core of art and design, and in fact, don't try, bypassing it altogether for objective and scientific truths that are entirely independent of us, and indispensable for professional standards to hold across the board. Still, the aesthetic impulse, intuitive and subjective, at the quick of art and design is undiminished and remains at full-strength for the fickle yet focused eye, whether the artist's or the critic's—"architects [are] concerned with appearances . . . someone has to be." David Watkin's visual aesthetics and corollaries of taste anticipate this, in *Morality and Architecture* (1977); it stirred things up decades earlier. And both then and now are fine as far as they go. That seeing and how things look overwhelm computers and calculating is obvious and irrefutable—it's hard to think otherwise. This is a promising start, but nothing close to the aesthetic enterprise I encourage in shape grammars comes up—do calculating, and art and design combine in some way, so that they're coadunate in a full-bodied aesthetic relationship? I guess that not posing the question is a tacit way of firmly answering, no—who would dare the opposite? Art and design go farther than computers and calculating—isn't that what art and design are for? Many take the lure of this dichotomy, secure in a lazy choice and too sated with artistic/scientific prejudice for change; at best, there are fashionable hats, separate and equal—one for computers and calculating and another one for art and design. A few look smart in both hats and enjoy them on different days, not seeing that they inform each other, being more productive as one instead of two. Wearing two hats one at a time omits individual differences and flaws that dissolve mutually. This is the how and the why of shape grammars, that won't take "no" as an answer. Shape grammars overtake computers to make designing things practicable in art, architecture, etc., when they exploit seeing that's rife with ambiguity, anywhere the eye may fall. It seems to me that without new perception and a meaningful way to explain it, there's no cause to worry about art and design—there's nothing more to say about pictures and poems, and whatever they evoke to each of us. And some add to this that the aesthetic impulse drives mathematical and scientific discovery, as well. Beauty isn't truth but helps to find it and to increase its sway; beauty is the locus of invention—try to get along without it. Nonetheless, this may be old fashioned, or so others say. Today, data and machine learning make scientific theories and physical laws unnecessary, and will replace them totally.

This doesn't matter for art and design—there Wilde's aesthetic formula and the critical spirit are paramount. What better way to understand the ins and outs of how all of this works than in visual calculating that allows in everything you and I see? This never halts—in one way or another, seeing moves on in its own way, indifferent to prior perception. Leon Battista Alberti (artist, architect and theorist, reputed author of the *Hypnerotomachia Poliphili*, keen cryptographer, humanist, linguist, mathematician, philosopher, poet, priest, raconteur and frequent dinner guest, the total Renaissance man—what wasn't he in/as a single soul?) kept to this aesthetic/perceptual measure six centuries ago, in his dialogue "Rings"—

> [The] eye is more powerful than anything, swifter, more worthy; what more can I say? It is such as to be the first, chief, king, like a god of human parts. Why else did the ancients consider God as something akin to an eye, seeing all things and distinguishing each separate one.

No one doubts that God is fickle, "seeing all things and distinguishing each separate one" differently at different times to suit him/herself in his/her own special way—it's God's world to re-create. We attribute to God what impresses us most about ourselves and of course, he/she reciprocates. God's way of seeing things and ours overlap—this goes both for the definite things that God/we pick out (the vast array of everyday perception in Coleridge's "primary" imagination) and for the flexibility he/she/we have to see again in another way (in the artistic/poetic imagination that Coleridge deems "secondary"). Nonetheless, not everyone welcomes the idea that artists and designers are gods, when there are mundane appellations that work as well. (It's easy to find names in a range of choice from polite to impolite.) Some tell me that it's brazenly aristocratic—capriciously changing your mind about what you see in order to design, without rolling up your sleeves and getting your hands dirty. But even if this is true, it's an aristocracy of perception and the senses, that invites universal participation at any time, from anyone who cares to join in and take part. Art and design aren't simply making (assembly, manufacturing, 3D printing, etc.)—they take, for example, more than craft that relies too often on numbing practice and monotonous repetition. Making doesn't explain readymades and *objets trouvés*, and what happens in photography and elsewhere, when pictures are taken and only made later, after finding time to look at them. Maybe all pictures/paintings are readymades. Nor do clever concepts worked out in advance cover art and design; nor does cognition that relies solely on descriptions and representations in discursive language and thought no matter how they're defined—compositionally with a recursive syntax in a vocabulary of atoms or word-like elements, or from set to set with mappings in an algebra of semantic units. Once concepts and the results of cognition are fixed and stored in memory for future use, to recite by rote, art and design end—scant if anything is left to/for the imagination. At most, there's a kind of visual etiquette to spell out good manners for looking, for example, in Gestalt laws, or that a gestalt switch (ambiguity) is limited to a pair of readings, maybe vibrating renderings in a duck/rabbit. You're taught/told what's correct (polite) to see, and that to forget is to fall into error. No, this spoils the trick—new perception (looking) is first and foremost in art and design, in ongoing descriptions and representations that aren't invariant, altering freely and flexibly in sync with rules in shape grammars, to inform making (from handicrafts, like knitting or lace, to painting and sculpture), and concepts and cognition, etc. in the many inventions of mind and hand. New perception is why pictures and poems overflow readily, to exceed intentions and expectations—there's open access to everything that hasn't been seen or experienced somehow. New perception is what rules are for—for visual calculating in shape grammars. The fabulous winged eye on the reverse side of Alberti's portrait medal, cast by Matteo di' Pasti around 1450, is a fitting image of this, in the force of its wings, and in its fantastic lightning bolts and sunbursts—as a symbol of vigorous creative activity that can't fail to illuminate in a flash of insight and a surge of imagination, to see anew in varied and expansive ways that excite and enrich the soul. The relentless pulse of the winged eye in untethered flight, wingbeat after wingbeat, is the insistent shock of new perception, forever on and on in shape grammars, resonant with the "embed-fuse cycle" for rules. The magical vitality in this comes out, as I show how rules work— the embed-fuse cycle lifts the winged eye in an extravagant census of what there is to see, to assimilate in order to re-create. The joy of flying diminishes the incumbent risks, terrible and sinister ones in creepy spying to coerce and control, and to crush the soul in trained reply, in a planned-out and neatly ordered, arithmetic world. Art and design break away and veer off, irrepressible in visual calculating; this much is sure in shape grammars—the embed-fuse cycle is unbounded, full of ambiguity and ready for change. The trick is to really look, so that things invariably feel different as rules are tried, not worried if this agrees with anything given in advance in axioms or data, or even etiquette. There's no rhyme or reason to it—visual calculating isn't thinking that's snared in rigid logic or mired in dismal statistics. The winged eye swerves this way and that, free in an inconstant and unknown path to create with no limits, "more powerful than anything . . . seeing all things and distinguishing each separate one." (It may seem ironic that the winged eye swerves to change what it sees—it's what atoms do blindly in the Lucretian universe, to collide and create new things. Still, a swerve is any movement without a definite, predetermined goal; there's naught to imply atoms or anything else. I guess it's the same for calculating/recursion—it's no big deal for symbols that are always atomic and don't alter when they combine, and it's OK for shapes that really aren't this way, neither atomic nor invariant in the embed-fuse cycle. And neatly, as luck would have it, calculating with shapes allows for calculating with symbols. Even so, there are some conspicuous asymmetries across dimensions, from i = 0 to i = 1—only atoms are distinguished for symbols/points in sets, with everything yet to see for shapes/lines in drawings. Such parallels are nice, but to be completely honest about it, I just like the feel of swerve, and how things change when they do. My willfulness jibes with the intuitive thrust of willful creativity in new perception, in the rush of wonder and delight in every swerve, unaccountable in arc and perfect in effect.) In art and design, there's more to see whenever I choose to look; the winged eye never touches ground—

Alberti's motto—QVID TVM—below the winged eye is sometimes rendered as "what next." (In addition, QVID TVM is "what then" and "so what." The latter may make fun of Alberti's illegitimate birth in light of his amazing feats in life—origins, including original intentions, aren't decisive. It's the same for seeing and visual calculating in shape grammars.) As ecphonesis, "what next" may betray the sheer exasperation of those who urge panoptic surveillance (spying)—at the impossibility of classifying everything we see and do, even with mountains of prior (past) data in superfast computers with vast memory. The road of sin and good intentions goes forever on and on. That's why it's prudent to purge what doesn't fit in—in righteous cause against uncommon perception and outlandish behavior that jolt and offend, the telltale signs of mental lapse, moral decay, and bad taste that bring nothing except misfortune and despair. For everyone's health and well-being, seeing and doing should be taught, prescribed in hard and fast tenets (visual analogies) for all to keep—in official art, properly sanctioned and unyielding to question. But QVID TVM is also for Alberti, perfect for his contrasting passions of counting and seeing—and in each, there's something unknown. On the one hand, "what next" is an arithmetic query about sequences and secret codes, to find a pattern in a network of numbers and relations. The goal is to predict from prior facts and findings—today, for example, with C. S. Peirce's abduction to guess instinctively in our peculiarly human way for scattered data, or with the statistical methods in machine learning to decide effectively for big data. Then on the other hand, "what next" is the aesthetic impulse to look again as if purely by chance, to see (observe) in a different way, so that whatever you find is a surprise, noticed before or not. The goal is to perceive in total ambiguity. Somehow, prediction and perception are largely at odds—even miles apart. The kind of analysis in rules (formulas and mappings) defined/discovered for counting is only dull censorship for seeing. Formulas and mappings, it seems, are impossible to keep as long as there's new perception (von Neumann limns this gap for visual analogies and the Rorschach test—the rote results and confident conjectures of the former fall short of the impulsive/personal quirks of the latter). There are no formulas or mappings for pictures and poems. This conclusion is routine for artists and critics alike, yet it may not mean what you think. Seeing is no less calculating than counting—both are recursive, and in fact, counting is a special case of seeing, that proves right for computers. But I've said this already, that calculating with shapes includes calculating with symbols. In what comes next, the swerve from counting (dimension $i = 0$) to seeing (dimension $i \geq 0$) shows how symbols and shapes are related, without recourse to representation—applying rules to add numbers and concatenate symbols in customary ways, and applying rules to see shapes and change them in expansive, strange and wonderful ways for endless pleasure and delight. This grand synthesis extends calculating past fancy (computers) to traverse imagination's magical realm—shapes beyond numbers and symbols. And this is the locus, too, for the many drawings I've used in "Seven Questions" and in my exhibits, that play active parts inside and at the end of sentences along with symbols and words. It's good to copy these drawings freehand as you read, and to intercalate between them in your own drawings and sketches in rich marginalia, in order to calculate from one shape to another, to be sure step by step of what you see. In visual calculating, seeing informs drawing in countless ways, so that things change whether they're embellished and revised by hand with new marks and lines, or varied just by looking. This is an integrated process that enfolds ambiguity in each and every step/swerve. Nothing is ever fixed; nothing important is taught—things of value alter as I go on. With rules, I do what I see. The test for calculating, both visual and not, isn't so much to encode what requires training and expert knowledge or to mimic everyday habits and elusive common sense—common sense and scientific opinion agree, that this entails mapping out the brain and its structure to mirror how the brain as in itself really works. Rather, the true test is to embrace insight and imagination, to find value in things extravagant and odd, not settled or known in prior experience—in shape grammars, this means looking at shapes in drawings and paintings, etc., in any of the untold ways they are as in themselves they really are not. That's the reason for visual calculating and what happens in shape grammars, whenever a rule applies in the embed-fuse cycle. This isn't finding descriptions and representations in order to calculate, but calculating in order to define descriptions and representations that are free to change with every rule I try, sometimes smoothly, and sometimes with jumps and breaks—continuous only in retrospect. In shape grammars, calculating relies on the eye (perceiving) to subsume mind (thinking) and hand (making). Without perception, thinking and making are idle, with scant to go on. Shape grammars are how to calculate when you don't have a clue what you're doing until you stop and are finally done, and even then, you can never be totally sure, because anything goes. Minus this kind of calculating, there's plenty that computers can't do—especially when it comes to art and design. The aesthetic impulse for looking underlies shape grammars; it's why they're special, and why they work for shapes in surprising ways, and for symbols, as well—to supersede counting as seeing and calculating meld, with the unlimited esemplastic power of imagination. This assays visual calculating in shape grammars against a necessary criterion for art and design. Imagination is a hard test, but I promise high marks in neat tricks. They unfold in wanton ambiguity, coadunate in "Seven Questions" and the ensuing three exhibits, and in their abundant notes.

—George Stiny

June 2021

# Acknowledgments

Many people helped with this book. Over the past decade, the members of my seminar (*salon*) in design and computation at MIT (4.581) have contributed enthusiastically with known questions framed in new ways, and with offbeat material and oblique perspectives to assimilate—a seminar of eight adds an extra 16 eyes that multiply and compound in neat ways year after year. It's been a diverse group—beginning graduate students, visiting ones, and ones returning time and again, and also established academics at sundry universities in the United States and from many places around the world. Then there were my myriad conversations with Lionel March and similarly in reprise, with Edith Ackerman and James Gips that reached their ends over the past few years—it seems that death is vast silence and lost friends, no longer there to care about what you've done. I miss them—in both ways it ("them") refers, as friends and for all they had to say. Nowadays, I feel their influence in the many new ways I re-create their words in fluid memory—this is the reason for ongoing conversation that always has special shortcuts, and restarts full-strength anywhere you wish and jumps wherever you want it to go, to provide the push and pull to traverse an expanding and ever-pulsing landscape in order to see things in an altered light. Christopher Isherwood is particularly good at describing the delight in such "rigamarole" that trades perfect (no noise) communication for luxuriant (noisy) understanding, for W. H. Auden and his group of friends—

> We were each other's ideal audience; nothing, not the slightest innuendo or the subtlest shade of meaning, was lost between us. A joke which, if I had been speaking to a stranger, would have taken five minutes to lead up to and elaborate and explain, could be conveyed by the faintest hint. . . . Our conversation would have been hardly intelligible to anyone who had happened to overhear it; it was a rigamarole of private slang, deliberate misquotations, bad puns, bits of parody, and preparatory school smut.

I try to get away with this sometimes, whenever I can in the way I answer questions. And of course, there's the Greek axis—Alexandros Haridis at MIT, and Sotirios Kotsopoulos at the National Technical University of Athens (NTUA) and also MIT. They read "Seven Questions" at various times, and offered invaluable suggestions and sound advice as it took shape in its present form. Sotirios tried many versions of "Seven Questions" in his classes at the Polytechnic, providing real-world feedback. Moreover, he insisted that I look at Auden's "Journal of an Airman." (Auden read another poem at Harvard in 1946—"Under Which Lyre: A Reactionary Poem for the Times"—in which the "sons of Hermes" and "Apollo's children" are locked in strife over ancient/idle tricks and advanced/useful technology. "Precocious Hermes" pits his lyre for tricks in pictures and poems against "pompous Apollo's" with its tenets for planning a neatly ordered, arithmetic world. This "fanatic" clash of half-brothers and two cultures produces new ratios that match with shapes and symbols, as thesis and antithesis—first for Hermes and Apollo, and then for pictures and plans, among other pairs that highlight the value of wanton perception outside the bounds of polite structure. Today, computers and data ignore Hermes to claim Apollo's victory, unaware of what's lost. But in shape grammars, there's a synthesis—full of tricks that subsume smug technology, Hermes runs rings around Apollo to exhaust proven order in the mischief and confusion in open-end ambiguity. This is straightforward when the dimension $i \geq 0$, and a tad harder when $i \neq 0$—the extra effort shows that lines, etc. can be used like points with no decrease in the embed-fuse cycle.) Alex did the drawings that take this book from words to something more to see, rendering my rough sketches in well-crafted lines and planes in order for these shapes to attract the eye and to hold its focus, as it flashes from here to there—to show how shape grammars work visually, in the best of all possible ways. And additionally, he put up with the excess of my drifting moods and the visual changes they invariably implied, as he made sure that everything repaid seeing time after time. Everything is the same throughout—nothing takes away the looking.

# SHAPE GRAMMARS: SEVEN QUESTIONS AND THEIR SHORT ANSWERS

## Q1.

What is a shape grammar? If you were to write one down on a piece of paper to show someone, what would it look like?

## A1.

A shape grammar is a way to calculate visually, entirely in terms of looking and what I see. In math and logic, and for computers, calculating is symbolic; that is to say, defined for indivisible units or atoms—for primitives like 0's and 1's or A's, B's, and C's. This is key in Alan Turing's famous 1936 essay on computable numbers, which is the gold standard against which all calculating is measured.[1] In "Turing machines," and equally for computers, calculating is essentially counting units and their combinations in this way or that. Shape grammars are off the scale—they allow for symbolic calculating as a special case, going beyond this for shapes that aren't made up of units, or analyzed/divided into parts of any kind. There are no underlying structures for shapes to put into descriptions and representations of them, or any way to count their parts—divisions and relationships in shapes depend on calculating, and vary as it unfolds. In shape grammars, units and parts result from rules and how they apply; units and parts aren't a prerequisite. Shapes are infused with ambiguity throughout—they're too changeable to describe once and for all. The willingness to make use of ambiguity freely and not to worry about it is key in visual art and design; otherwise, shapes and pictures would always stay the same—inert, fixed, and dead in every respect. There wouldn't be any reason to look at them even for a single time; simply describing them, counting out their units and parts, would be ample for memory and use. In visual calculating in shape grammars, rules are for seeing and doing; rules exploit ambiguity in order to supersede counting. There's the kind of "negative capability" that John Keats prizes in William Shakespeare—the knack for ambiguity, contradiction, discontinuity, uncertainty, and luxuriant, open-ended change, that remains content with half-knowledge, unperturbed when things are unsettled or impossible to resolve. Some of the mathematical details for how shape grammars do this are sketched later in Exhibit 1. This runs counter to my emphasis on visual calculating strictly as seeing, but it's another convincing way to show that shape grammars really work, without any irritating bugs or snags that stop their use unexpectedly or cause them to seize up—it's the proof that many insist on seeing. Visual evidence is given both in Exhibit 2 and in Exhibit 3. Ambiguity and how to take advantage of it in art and design is the focus of the former, and different ways to use shape grammars in teaching with schemas for rules are sketched in the latter—these exhibits highlight the accessibility and fluidity of calculating with shapes. It's amazing how effortless this can feel when there are rules to see.

The visual basis of shape grammars makes them easy to define—any pair of shapes is all it takes to make a rule that works with any other rules. It's OK to draw these shapes by hand, using pencil and paper—any two drawings, whatever they are, are a rule

drawing1 → drawing2

that I can use recursively to calculate. Of course, "drawing1" and "drawing2" are already actual drawings in themselves that I can trace to include either full letters or their parts and segments as I please. But I can also take "drawing1" and "drawing2" as open variables (abstract symbols) in the usual way. Then drawing1 might be a square with four edges/sides—that's easy enough, it's how I was taught in school—and drawing2, the square in drawing1 with another square that's half its size inscribed inside it, vertices to edges at midpoints, four lines twice equals eight lines in all. The rule is a cinch to draw; it looks like this

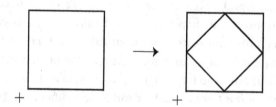

However, what I've said isn't ironclad. Drawing2 with the parts I've described might also be four triangles, two separate pairs of pentagons in six ways, double hexagons, letters of the alphabet—maybe there are myriad K's and k's—or anything else I can find.

Shapes are what I see—what I "embed" in them—not merely what I think I've drawn or what someone else tells me they're supposed to be. I can put into them anything I choose, whenever I want. One way to explain this is that shapes have indefinitely many symbolic descriptions—but for any given shape right now, I have to calculate to figure out which descriptions work the best. The shape is thus and such with distinct parts and relationships if I stop, although this structure isn't fixed if I decide to go on calculating. Pointing to my rule is probably a good way to talk about it, without trying to decide what's in drawing1 or drawing2, or naming these drawings and their parts in terms of the shapes I know (squares and triangles, etc.) or the ones I like that unfailingly seem to look right. If I want to play it safe, the most I can do is to use two more shapes as variables, indefinite pronouns that give me something definite to say

this → that

as I wave my arms and hands around vigorously, and jump up and down in my effort to point out what's there. Ostension seems to work in a largely visual rite that doesn't rely too much on words. Somehow, we're able to figure it out. My generic rule

drawing1 → drawing2

applies to a given shape, also rendered in a drawing, to change it. If part of this drawing looks like drawing1 in some particular way—tracing paper is a real help here—then I can erase it or take it away to add in another part that looks like drawing2 in exactly the same way. The known part is to the new part as drawing1 is to drawing2, in an analogical relationship with a rule of three that produces visual results from shapes—say, a linear transformation, maybe a projective distortion or more narrowly, a similarity transformation or a Euclidean isometry. There's no denying that isometries and similarities of a given shape look like it—at most, they vary in location, orientation, reflection, and size, so that angles in lines are invariant. This may not go for projective transformations, etc. In essence, I can embed a copy of drawing1 in my drawing in order to replace it with a corresponding copy of drawing2 that solves a given analogy—this is what it usually means to follow a rule in a shape grammar, seeing (embedding), and doing by erasing (subtracting) and drawing (adding). Of course, there's always seeing—that's constant and never lost in visual calculating—but sometimes, the details of doing may vary—maybe I scribble at will, blot with a sponge full of color and dye, or throw paint in some special way that's my own. There are myriads of ways to make shapes, whether I know what they'll be or not. The important idea to grasp in all of this is that the rule

copy(drawing1) → copy(drawing2)

is the same as the rule I defined for drawing1 and drawing2, but tailored just so, to match my drawing or some part that's a piece of it. I normally assume that the rule drawing1 → drawing2 is the name of an equivalence class, for example, when groups of transformations are used to determine copies. Whether this is hard and fast or not, however, has yet to be explored—it's something to think about. (There are Boolean expressions for how rules work in this way, framed in terms of the part relation ≤, and the operations of sum +, product ·, and difference –. All of these are defined rigorously in terms of embedding. Embedding is at the quick of visual calculating in shape grammars—without it calculating reverts to symbols and Turing machines. This makes the proof straightforward, that Turing machines are a special case of shape grammars.) Maybe I calculate like this

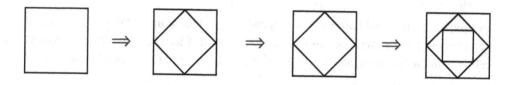

In this three-step process, it's easy to see that adding is seamless. Everything "fuses" as one, as I change my original shape to draw a new one—the small square is drawn twice in the third shape, but there's only one small square to see, and that's what matters. What you see is what you get—parts meld and boundaries blur and fade away. The next time I try a rule, what I've added or drawn in may have disappeared; there's no memory of it. Yes, this part is visible—I can see it if I have rules to look for it (memory, like structure, is lodged in rules and their use, not in shapes)—but untold other parts may pop out, too, those neat surprises I can't ignore, that appear as I go on trying rules. It doesn't take much for visual calculating. Drawing according to drawings in rules is pretty much all there is to it—it's simply doing what I see, with no intervening structure or anything else to block the way. One of my research students took a graduate subject with a computer scientist and linguist who was sure that drawings alone weren't enough to define rules and to use them—symbols and numerical values are necessary to represent shapes in order to calculate. But drawings are all I need to get the forms I want, and many other, extravagant things as well. Shape grammars let me exceed any intentions or plans I may have had to start; they work for what I see here and now, cool to definite ends and goals, and indifferent to what I think I've done—this history is easy to invent and reinvent as rules are tried. One of the really nice things about shape grammars that's entailed in this is their visual inclusiveness—they allow for universal participation no matter what anyone sees. There isn't anything to know in advance to see more, whether starting out or going on. Prior knowledge (what shapes and parts there are, and how they're described or represented) isn't required—there's nothing to learn or to keep in memory, or that's hidden from view, censored or left out, to curb what there is to see. Everyone has the same chance to take part and to join in, at any place or any time, with rules that find what they see, no matter if this is commonplace or strange, from memory or totally out of the blue. No one is privileged when it comes to putting things into shapes, resolving their parts, or seeing and saying what these are. There's no truth or right and wrong when it comes to shapes and any of the varied ways they may appear. In fact in shape grammars, participation limited to one alone, without outside influence, is no different than participation for many interacting cooperatively or competitively, in concert or at odds. With rules, one sees in the same way as many, and many see the same as one. Everything is wide open for individuality, eccentricity, and nonconformity, and for community and solidarity in multiplicity and diversity. The trick to all of this is embedding, in shapes that fuse every time any rule is tried.

## Q2.

How far can shape grammars go in describing forms—can they describe anything? What are they good at, and what do they struggle with?

## A2.

Shape grammars let me calculate anything I can draw on a piece of paper or I can make in three dimensions. In this sense, there aren't any limits as to the forms shape grammars can be used to describe—their sweep is unbounded. And this comes with an added bonus. Shape grammars also let me describe what I see, for example, in terms of symmetry or in a parsing process that picks out (embeds) meaningful/useful parts and shows how they're related to one another in different structures—graphs, hierarchies (trees), topologies, etc. Identity rules that look like this

$x \rightarrow x$

where drawing1 and drawing2 are exactly the same, are particularly good for symmetry and the kind of parsing I have in mind. Identities $x \rightarrow x$ may seem pretty useless—at least most of my students think so—because they don't change anything when they apply. (My word processor agrees wholeheartedly with my students. It's certain the right-hand x is a mistake, because it repeats/copies the left-hand x in close proximity, without enough distance for individuality. For computers, identities are typos. Crucially, though, identities in computers produce infinite loops for single symbols and fixed words, and that's something dire to be avoided.) But embedding goes on all the same for identities to show the way visual calculating works in shape grammars. Shape grammars let me use anything I can see, and identities are the rules that let me pick this out without altering anything. Seeing is enough for calculating in many nifty ways— it's amazing how much there is to see and what I can do, simply by looking. (Identities are fun to play around with in terms of how x is described as drawing1 and drawing2. Shapes in identities may seem to be really different things even when they're not—and these differences can be used when I calculate. That's the beauty of shape grammars. The shape

in the left-hand side of the identity

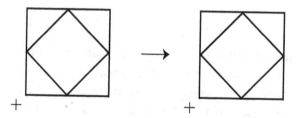

may be two squares

 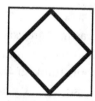

and in the right-hand side, four triangles

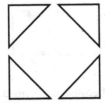

It's also the same for the square

In the left-hand side of the identity

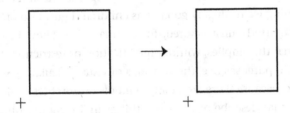

it may be four edges

and in the right-hand side, four corners

In both cases, there's an identity because the sum of the parts in its left-hand side is equal to the sum of the parts in its right-hand side—parts aren't divided in rules but always fuse. And it's an easy corollary that the number of parts in the left-hand side may or may not be the same as the number of parts in the right-hand side—any number of parts will do for either side. More accurately, though, the shape in an identity is unanalyzed in both of its sides—what it is depends on how the identity is used to calculate, and this may change as calculating goes on. In fact, the same goes for any rule—the shapes given to define a rule are always unanalyzed. This is the reason shape grammars are so powerful in art and design; shapes can change what they are time and again—anything can be different than it seems, without an end to calculating.) And then there's what's hard. It isn't so much that shape grammars struggle with some things, it's more that people seem to struggle with shape grammars, to say they're tricky to use and a formalist aberration—even when it only takes seeing and drawing for shapes and rules, something we're already good at as children. I guess no one expects shapes or anything to be utterly unstructured, for calculating to allow in visual ambiguity and everything that this implies, without qualification or restriction. I just mentioned that to some in computer science, the very idea of visual calculating seems impossible, silly and *ubuesque*, because there aren't any symbols or numbers—calculating is exclusively 0's and 1's that describe or represent things in Turing machines. Others in art and design find visual calculating twisted, in need of a moral compass—seeing in too many different ways lacks sincerity and authenticity; it's shameful and can't be trusted. Visual calculating raises questions that are justly left to creative artists and designers in their personal work and private (secret) practices. What gives you the right to how others see, or to anyone's soul? The quick of creative activity in art and design is totally off limits. (My parents missed their chance to name me Frank. With my obligatory middle name Nicholas, I'd be Frank N. Stiny—so much for truth and wisdom in legal/permanent names given at birth, to prefigure later outcomes. Of course, Mary Shelley didn't anticipate this for her modern Prometheus—how could she? My surname wasn't anyone's name in the 19th century—my grandfather invented it in Denver, in a blaze of awareness and foresight, in a fresh start as a Greek immigrant a hundred years later.

Such tricks in infantile puns are below the threshold of serious art, jokes only to pread-
olescents and childish nerds, but maybe Shakespeare and his ilk are exceptions. Mess-
ing around with names and sounds can be inane and not; in my pun, Frank N. Stiny
makes a key point—"perhaps the component parts of a creature [name or better, shape]
might be manufactured, brought together [fused], and endued with vital warmth."[2]
Why not wait and see what happens once things are reanimated and in motion—
names bestow new life and élan. Frankenstein's monster doesn't have a name, but
enjoys a nomenclative tie to artificial intelligence in calculating and the Analytical
Engine. Shelley and her husband were ensconced in Geneva when *Frankenstein* was
first rendered in words, guests of Lord Byron—George Gordon Byron, 6th Baron of
Byron—who fathered Charles Babbage's protégé Ada Lovelace—Ada King, Countess
of Lovelace, born Augusta Ada Byron, Lady Byron. Names abound among peers, and
change for performing artists, writers, and even mathematicians. In French, there's
plenty of scope in *noms de plume* and *noms de guerre*. Aliases, pseudonyms, and masks
abound for children, and this magic turns tricks for clairvoyants, divines, politicians,
prostitutes, swindlers, and thieves. Native Americans change names to describe major
accomplishments and mark key events throughout life, and less formally, nicknames
may trace personal histories in the chaotic flux of comings and goings in daily affairs.
There's a lot in names—at the very least, as an evolving record of how things change.
In shape grammars, prior names alter every time another rule is tried, to describe what
calculating does here and now to influence what went on before, and how the present
invariably reanimates/reconfigures the past to infuse new meaning and value in shapes
and their parts in surprising ways. Settled names/descriptions are secure at last only
after calculating ends, when seeing is finally over with no intention of going on. Then
shapes are classified and docketed—essentially fixed and dead, monotonously tallied
one after the other in the gridironed graveyard of rote perception. Topologies defined
retrospectively show how this works, to ensure continuity for rules in general.[3] Then
rules are expressed in symbols like this

$$x \rightarrow y$$

and for identities $x \rightarrow x$, when $x = y$.) There's no reason to think about visual calculating—it can't be done, and should never be tried. Maybe so, but visual calculating
is what shape grammars do. I've tried them out time and again, and they're up to
the job—and much more, too, that's unexpected, strange, and extravagant. In shape
grammars, rules work outside the standard precincts of possibility in logic (combinatorial play) and ethics. True and false and right and wrong aren't meant to exceed the
acknowledged constraints of 0's and 1's in symbolic calculating—this is the eternal

source of their rigor and strength. But excess is inevitable for seeing that's rife with ambiguity, when aesthetics (perception) is prized above all. (This may seem frivolous and irresponsible with the endless problems that plague today's harsh and unjust world. Still, aesthetics does necessary spadework for logic and ethics, as it charts its own way in art and design, driven by feeling/sentiment and taste to new things.) Excess is what visual calculating in shape grammars is for—to put shapes in motion as you go by what you see and not by what you already know. It's following rules to change shapes even if this is just by looking, without describing them somehow before you start in terms of given parts and their relationships in fixed structures. This supersedes calculating as usual, in logic and ethics, and throughout reason and thought, where descriptions/representations are prerequisite everywhere. Aesthetics makes a real difference; seeing alters everything—no prerequisites required.

**Q3.**

What inspired your seminal research into shape grammars in the early 1970s? What fields/interests was it coming from, and who was it for?

**A3.**

I've always been interested in what visual artists and designers see and do, and this includes my own work. What is it about pictures that makes them so engaging, and forever worth another look, to see what you've missed that may add to, take away from, or even replace what you initially thought was there? Inventing shape grammars was my way of finding out. On the one hand, it seems OK today to use calculating to help explain things, so shape grammars are a good place to start. And on the other hand, shape grammars expand calculating in an unexpected way with embedding, and thereby include art and design. Calculating with shape grammars says something about what artists and designers see and do, and reversing this, what artists and designers see and do says something truly new about what calculating should be, and why shape grammars are necessary to embed and fuse shapes and their parts, as rules apply in a process that's entirely open-ended. Turing and John von Neumann invented the kind of symbolic calculating taken for granted in computers, and used all over the place today. But maybe there are real alternatives. Von Neumann suggests as much when he compares the "visual analogies" needed for symbolic calculating (my "spatial relations" like a triangle with three edges or a square in a square, and other descriptions and representations like this that are built up in terms of given units) and what happens if you try to describe putting something into a picture or a Rorschach test, in a personal way that's intuitive, fickle, and an ongoing surprise—

> In addition, the ability to recognize triangles is just an infinitesimal fraction of the analogies you can visually recognize in geometry, which in turn is an infinitesimal fraction of all the visual analogies you can recognize, each of which you can still describe. But with respect to the whole visual machinery of interpreting a picture, of putting something into a picture, we get into domains which you certainly cannot describe in those terms. Everybody will put an interpretation into a Rorschach test, but what interpretation he[/she] puts into it is a function of his[/her] whole personality and his[/her] whole previous history, and this is supposed to be a very good method of making inferences as to what kind of a person he[/she] is.[4]

What good is calculating or using computers, if every shape is a Rorschach test with no visual analogy to represent it, when deciding what to describe in a picture never ends— how does this test calculating, and trace its limits in order to extend them? What would calculating be like if Turing and von Neumann were painters and not logicians, or is this too farfetched to heed? Actually, it's a pretty good way to look at shape grammars— visual calculating opens uncharted vistas via art and design. This may be something of a shock; without shape grammars, it seems implausible that art and design are calculating. But seeing how shape grammars work proves this equivalence. Instead of keeping subjects apart, I like to put them together, so that prior divisions simply disappear with no memory or trace. I'm not keen on multiple cultures—paradigmatically, two, art and science (calculating), where what makes sense in the one is obscure or risible in the other, both foreign and unwelcome. Some try to balance such cultures and rival points of view by sporting different hats on different days—yet even when this works, it doesn't justify hats individually, settle their differences, excuse their flaws, or bring them any closer together to adapt reciprocally and change. I guess wearing different hats is a good way of showing your mental agility, and that you're generous and not one-sided—a modern Renaissance man, a universal Turing machine. But I'm drawn to circles more than polygons that add needless discontinuities (vertices) and distinctions, and figures with unbridgeable breaks and gaps—all hats are the same on a circle, each melds into the others with no distinguishing names. It may be worth trying to explain why, even if it's more of a spiritual feeling than anything I can spell out fully in words. Others revere the gap (rent) between counting and seeing that shape grammars bridge (mend) out of genuine feeling, too—no reason why. For the mathematician Gian-Carlo Rota, there's combinatorics (counting) and discretely (discreetly), phenomenology (seeing), although he doubts the standard divisions in math and with Stanislaw Ulam, bets on a key aspect of shape grammars in calculating—seeing *as*, as a Rorschach test.[5] (Ulam played a large part in 20th century math and physics; von Neumann and he invented cellular automata, usually defined for patterns and forms in square grids—John Conway's "The Game of Life"—yet entirely symbolic.) Sometimes, mathematicians are complicated and easy to like—asked about his two hats, Rota simply replies, "I am that way." Is this answer any better than mine, that it feels right for hats—why not counting and seeing—to meld and disappear? Separate hats work like visual analogies—once they're on, they're inevitably invidious, but not really, in a more inclusive hat with a subtle/ironic point of view. A rack full of hats won't do—you can only try one on at a time, that may or may not suit the occasion with no time to try another one. Special hats don't alter when you wear them on successive days. It's combinatorics on Monday and phenomenology on Tuesday, so that counting and seeing go on as usual—neither influencing the other, from one day to the next. Maybe it's OK to keep hats just the way they are, but maybe there's more in what they're not? No doubt, hats are swapped hesitantly—things that go smoothly in one may rub in another. Monday's hat may be worrisome or embarrassing, if it's out of style or new, or you fancy Tuesday's hat as much or even more, with no way to ease this anxiety. A well-worn hat feels nicer than the rest; it puts everything right, with pleasing sights in easy view. Maybe it's better to enjoy crazy hats without putting them on—pretend they fit or leave them on the rack. But then, how do you change your point of view? Why are there alternative hats anyway, and where do they come from? I don't swap hats to see in different ways—I can do this on the fly in shape grammars, and make new hats as I go on, in visual analogies that change in terms of what I see. Still, I like to try hats in fluid succession to see how they differ and go together in a pleasant haze, most Saturdays at Salmagundi in Boston's Jamaica Plain or on Salem Street near Polcari's Coffee in the North End. It's perfect splendor as boundaries blur and dissolve—hats assimilate and fuse. Diverse shapes and varied colors in feathers and bows combine freely and commingle to modify each other mutually as one, coadunate in surprising ways. It's a lot like trying rules in shape grammars to see how shapes change as I calculate. The crux of my seminar (*salon*) at MIT, for shape grammars in art and design, is to trace the locus of visual calculating in the arts, literature, criticism, etc.—to grow many things in shape grammars that seem unlikely to ever root and thrive. Some of these are widely known and some are more obscure—plenty of cool stuff eludes expert approbation and popular notice. It's probably right to start in the arts. Shape grammars include Owen Barfield's "participation" and "figuration" at the quick of poetic experience, in the ongoing use of rules that embed and fuse shapes. Participation (using rules to see, fully in sync with the universal participation I stressed at the end of A1) is "original" when figuration (embedding) is immediate, and it doesn't worry us; this is the primary

task of the "primitive mind." And participation is "final" in the self-conscious/imagina-
tive re-creation/reaffirmation of "idols" or "collective representations" conceived in
prior figuration—effectively, the visual analogies we use to curb the overwhelming
exuberance of original participation, and to ensure coherent experience.[6] (The commu-
nion-like ritual in final participation unfolds in the neat math in note 3, at least for
shapes—graphs, topologies, and other visual analogies are defined in retrospect, to
ensure a kind of narrative continuity that shifts whenever rules are tried.[7]) Going
from original participation (embedding) to final participation (representation in visual
analogies, etc.) tracks seeing as its focus narrows, although this isn't ever for sure. Participation in figuration never ends—there's always plenty of room in pictures and poems
to see anew. Embedding goes on freely for everyone, even at the risk of "chaos and
inanity" in a "fantastically hideous world"—"We should remember this, when we see
pictures of a dog with six legs emerging from a vegetable marrow [shape] or a woman
with a motor-bicycle substituted for her left breast."[8] (Barfield uses the expression "saving the appearances" in its original sense, in the title of his book and throughout its
chapters. No matter how distant descriptions/hypotheses for the same things/phenom-
ena may be, they're equivalent—things are what they seem, our impressions are always
right. This OK's all there is, dividing shapes into parts; yet exactly what is, isn't a question to decide. Shapes aren't Galilean, where scientific truth is definite and final, and
something to docket and defend. Artistic/figurative truth relies on Epicurean plurality,
not Ockham's Razor. Seeing, and likewise embedding, takes in whatever it wants,
unthreatened by more; seeing saves the appearances in myriad ways.) Of course, there's
no reason to be crazy or far-out, or even to try mild alternatives. Barfield is conservative
and religious, with a profound sense of obligation to origins and continuity, and a sincere respect for objective nature in particular things, for example, a rainbow or a thrush
singing in the play of leafy idols.[9] This is cause enough to worry about rules in shape
grammars, and to control and regulate how they're used in ordinary experience.
Barfield gives three reasons for caution that extend readily to rules. (1) Rules eschew
collective representations wantonly, disfiguring nature without regret as they do, to
save the appearances in misshapen, six-legged dogs chasing bikes in alarming places;
(2) rules, grounded only in embedding, "revert [regress] to original participation
(which is the goal of pantheism, of mediumism and of much so-called occultism)," and
Barfield is quick to reject this outright—"it is no part of the object of [my] book to
advocate a return to original participation."[10] But (3) rules work just the same—
unperturbed, they calculate shamelessly with whatever "[original] participation renders . . . less predictable and less calculable."[11] Many who sport special hats fight shy of

rules, as well. Shape grammars allow in too much even at a glance, that's impossible to handle in a rush of possibilities—the eye arcs far and fine. No matter, shape grammars are useless in art and design, merely idle math when seeing wanes, enmeshed in social relevance—their fate was sealed in architecture, as they were invented. Serge Chermayeff (he was Christopher Alexander's PhD supervisor at Harvard) called for a new kind of practice—"Let us not make shapes: let us solve problems."[12] And Leslie Martin (Lionel March's champion at Cambridge) urged much the same—forms should be "'built to a purpose' and thought out rather than drawn."[13] This turns away from delight (beauty) in the Vitruvian canon to focus on scientific procedure and careful research. The key to calculating in art and design, however, isn't to ground it in problem solving and thinking, but to ensure it's visual with shapes and drawing—that's the go of rules in shape grammars, and their how and why. This highlights delight, as it gives art and design full sway to add in science at any time—explicitly, when shape grammars align with description rules, maybe in visual analogies for practical things, and in implicit ways, with weights and their scattered algebras.[14] (There's a nice example of weights in A4, for fifty shades of grey and their endless successors.) The three categories in the Vitruvian canon interact in shape grammars in an easy mix—coadunate, to foster reciprocal links in an ambiguous/creative process. The swerve of firmness (physical performance) and commodity (utility) to science and proven results is scant reason to forfeit delight in art and design. John Ruskin puts delight ahead of firmness and commodity in *The Seven Lamps of Architecture*—the uncanny, numerical congruence with my own seven questions and their answers may be telling. Firmness and commodity support building to a definite purpose with positive aims and goals, but lacking the aesthetic third (decoration and ornament) "above and beyond . . . common use," there's no architecture—so much the worse for problem solving and thinking.[15] It's only delight that turns building into architecture. Yet even so, the "architectural scientist" takes the objective view in another tack that goes elsewhere, rejecting fickle "appearances" in its course—superficial drapery, decoration, and ornament—to expose "'naked' [invariant] built form with all articulation, openings and decoration ignored [stripped away]."[16] This is at once reasonable and an effective appeal to convenience and utility; it facilitates "encoding [describing and representing] built forms . . . so that all possibilities can be counted." These are the central facts and figures underlying the analysis and classification that delimit and direct productive problem solving. Buildings have to work, but must they be stark and bare, with no cosmetic allure, glamour, vibrancy, or visual appeal? What happened to architecture and delight—aren't they calculating, too? No doubt, building is crucial in our diseased and disordered times that beg for experts (architects and urban designers) with scientific hats to solve pressing

problems and to think out remedies to urgent needs, for a range of choice in some kind
of design space that's been carefully defined and settled rigorously in prior research,
and tested and re-tested in the laboratory for use, reliability, and success in the real
world. Maybe delight has a salutary purpose after all—as an objective measure of choice
(optimization) for built forms. This new relationship updates Vitruvius after two mil-
lennia, to face the endless challenges and untold perils of modern practice in a handy-
dandy, no-nonsense, get-down-to-business formula; it subsumes delight in firmness
and commodity, with no autonomy of its own—

firmness + commodity = delight

This may include Chermayeff and Martin, and Alexander in the hierarchical analysis/
synthesis of form in graphs and pattern language, and March in configurational studies
for polyominoes and his truly amazing tally of abstract building plans; it lets experts
succeed confidently in a pat equality, trained in advance to know what to do.[17] Still,
this is sorely incomplete, minus appearances that enable personal meaning and pleasure, maybe in idols, to use Barfield's apposite term, but usually inconstant, and utterly
indifferent to representations, when seeing swerves wildly. Without seeing, meaning
is lost in ambiguity, bereft of feeling and value in participation and figuration—there's
no perception to distinguish (embed) things in parts or to hold (fuse) things together,
undifferentiated and whole. Vitruvian delight in appearances is vital; it's the nub of
visual calculating in shape grammars—no hats required. Today, architects, artists and
designers, engineers, historians, scientists, and others bound to professional standards
continue to fret in three, in parallel with Barfield—(1) embedding is too licentious,
or likewise, generous and inclusive, flouting fixed divisions and tried and true norms
for who knows what; (2) shape grammars track too close to genius/folly at its source,
embracing strange and magical devices (rules) that are much better left alone; and (3)
it seems that visual calculating works in theory, who's to say, but this isn't the way it's
supposed to be for computers, without symbols, shape grammars are trivial and won't
do. Shape grammars run art and design through calculating to make it visual, thereby
opening up art and design—in this way, they tie in other subjects and blur established
boundaries capriciously. Why not try a different method with odd ways of going on,
to see what surprises this holds? Only blind prejudice or a mathematician's love of
rigor—usually a good thing—would keep to fixed (standard) divisions, with this hat or
that one to limit seeing and doing, when new divisions are obvious and true. In fact,
participation in the way I talked about it in A1 already spells the end of hats. (This also
goes for Barfield's original participation, even if the goal in final participation is to con-
secrate hats and their contents in idols. In both, the key is figuration, and in it, there's

art and design—originally, as things are defined/resolved and change, and finally, in alternative histories of this recounted from memory or reproduced in other modes of figuration, anything goes from true stories to fake news, fantastic myths, and dark tales to varied graphs and topologies.) Seeing exceeds what hats allow separately, and in succession. In shape grammars, rules provide the impulse to see things anew. What's settled and known doesn't stay that way for long—to bring in Ezra Pound's formula in *ABC of Reading* in an effortless conjunction that adds an inclusive term (see the second parentheses in A4 for more on this), it's changed and charged with meaning in an ongoing and totally open-ended process.

In my student days at MIT, I signed up for Marvin Minsky's "Theory of Symbol Manipulation and Heuristic Programming"—or a close concatenation—to find out what artificial intelligence (AI) was all about, from one of its inventors.[18] I remember enjoying this a lot, as I worked on symbolic descriptions of pliers and similar hand tools—in regular expressions for finite state machines. This was a big success, in spite of what my method left out; there was more to see than I could say in symbols. Rumor had it that Minsky, with his odd sense of fun, played around with silverware at the kitchen table. The game was to combine knives, forks, and spoons in stable designs that were of ever-increasing complexity. Breakfast was a real challenge—so much for intelligence. But looking at Minsky's game and how designs are made one utensil at a time reveals a lot; it's a kind of calculating, with spatial relations (visual analogies) given for rules—it's what I do comprehensively with spatial relations in "Kindergarten Grammars" for Frederick Froebel's building blocks and their individual symmetries.[19] In fact, von Neumann encourages this in computers—descriptions (visual analogies and other kinds of data) and programs (rules and instructions) needn't be distinguished. (Shapes and rules are this way, too—each shape includes indefinitely many rules, as the sum of their two sides. Sometimes shapes are made up of points only, and correspond to spatial relations. Then, the results are finite—for two units, points or shapes, there are as many as nine distinct rules, including an identity.) And there's a little seeing in Minsky's game, as well; it probably goes better in shape grammars than in AI. Fast enough, though, spatial relations fail, whether in shape grammars or not; they fall short of what's expected in art and design. Trying spatial relations out is a useful exercise, an experiment with pedagogical impact in the kindergarten and in the design studio—it works but not completely, and it shows why.[20] And maybe that's what I'm getting at—whatever it tries, AI only does so much. Even with Minsky as a pied piper, promising ever more and more, symbol manipulation hits a limit in von Neumann's pictures and Rorschach test—it relies on descriptions and representations such as visual analogies in place of things that aren't fixed, when aspects, features, parts, and properties are all up for grabs. The incumbent abstraction leads straight to a migraine that only shape grammars seem able to cure, with rules exactly in sync with eye (seeing) and hand (doing). Shape grammars abandon symbols for shapes—and along with symbols, goes Minsky's AI. It's true that AI influenced shape grammars in important ways—hand tools were a promising start—but AI stops before calculating in art and design really begins. (John McCarthy, Minsky's double in AI, was a grad student in math at Caltech in 1948, when von Neumann was there at the first Hixon Symposium to talk about the theory and organization of automata, and how visual analogies probe the limits of calculating in a picture or the Rorschach test.[21] McCarthy's later approach to AI in terms of symbol processing is included in von Neumann's visual analogies—but maybe all of AI is this way, even with data and learning. McCarthy relies on constructive/ parametric descriptions and representations for legends or picture captions that tell you what to see; they combine units in terms of intentions and goals, in some range of choice—not unlike spatial relations for Minsky's knives, forks, and spoons, in an array of utensils laid out in stable pairs, etc. And McCarthy has a nice example of his own, expressed in a few words that aren't hard to work out—"a vertical line, connected at its middle, to a horizontal line segment going to the right."[22] This defines two opposing cantilevers, for the leading point x of the horizontal segment, first at 0 and then reflected at 1, and varied crosses in the open interval, one with bilateral symmetry when x is at 1/2, and all of the others reflected across this axis, between 0 and 1/2, and 1/2 and 1. The three cases for the values 0, 1/2, and 1 are enough to illustrate what's possible. Each of the shapes in this trio is two lines in a visual analogy/spatial relation, in a set or another synoptic description that includes everything there is from 0 to 1—

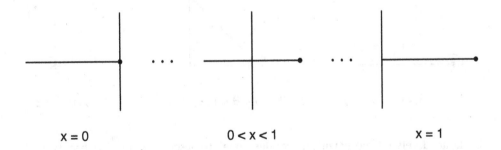

$x = 0$   $0 < x < 1$   $x = 1$

A century earlier, in *The Kinematics of Machinery: Outlines of a Theory of Machines*, Franz Reuleaux used the same method to classify the forms of a nut and bolt in terms of the pitch-angle θ of the screw—the angle of the slope or tangent for the thread distance

and the bolt circumference. The twisting pairs with helical motion are flanked at 0 by
a turning pair with rotary motion, an axle and wheel, and at π/2 by a sliding pair with
translational motion along a linear track.[23] For Reuleaux, it goes like so—

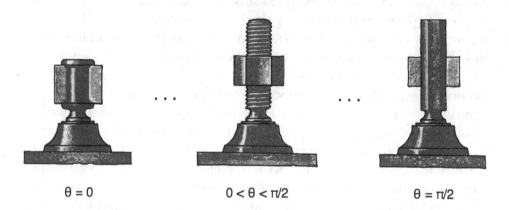

θ = 0                      0 < θ < π/2                       θ = π/2

This suggests a continuous transition starting with a pleasant stroll in the park on a
midsummer day, on level ground or a horizontal surface, where amusing conversation
is effortless and carefree; through heart-pounding climbs up inclined planes on steeply
rising diagonals, where labored breathing takes over with no respite; to a hair-raising,
heart-stopping ascent up a perpendicular, or a perfectly vertical wall, in an exhilarating,
death-defying climb—

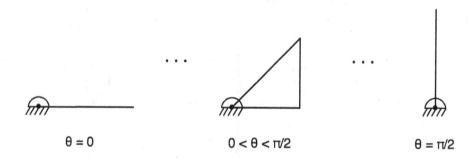

θ = 0                      0 < θ < π/2                       θ = π/2

Looks are deceiving, along with the stories we tell to describe them—like beauty, looks
run skin deep, superficial and fleeting, only in the eye of the beholder. This may obscure
timeless structures shared by different things. The parametric structure of McCarthy's
two lines, one moving with respect to the other, is exactly the same for Reuleaux's nuts

and bolts, and the corresponding inclined planes—no matter what I say about them. The mapping from x to θ is in the formula $f(x) = \pi x/2$. This isn't much of a result, and may be a letdown—but it's telling, that there's no magic in McCarthy's AI, just ordinary mechanical relations, kin to twisting and sliding pairs. It seems that AI is scarcely more than parametric definition, a super-hyped version of building information modeling that's known commercially as BIM, in architecture, construction, and engineering.[24] For all I know, McCarthy is right—intelligence is BIM with its parametric descriptions, object-oriented data structures, standards and measures, drag-and-drop format, etc., maybe on steroids from logic and statistics or some other programming PED. But this isn't seeing that thrives luxuriantly in open-ended ambiguity, and that soon withers and dies without it. McCarthy had a catchy slogan or jingle for his own personal style of AI/BIM, to chart the course for the research lab he started at Stanford—romantically, in letters/symbols, it spells SAIL. The salesmanship was pervasive, as the enterprise grew and grew in influence, renown, and size—

description and not merely discrimination

The double d's strike a pleasant balance left and right, a mnemonic that reverberates from pole to pole in this heuristic and guide, to tie seeing firmly to structure. Visual analogies are given for shapes, in order to recite in words how to draw them and how to recognize them, communicating over the phone with scant to see, and no time for talk, lies, and gossip. This takes away the looking—there's no need to decide what's there, or any chance to change your mind about what this is, or to elaborate and embellish as you please. Everything is constant, within some range of choice. It works the same in computer code, computer to computer, but not in shape grammars—they greet description versus discrimination with a listless yawn.[25] Embedding is the sine qua non of discrimination, or to put it in a slogan

perception without description

There's plenty of room now, to change your mind about what you see, time and again just because you feel like it. Of course, other relations do the trick; McCarthy's is also OK, in place of the prepositional connective "without"—

perception and not merely description

But this seems flat—no matter how often I repeat it, there's nothing to inspire change, merely meh. Words are odd things; maybe it's their sound, or the way they look on a page—p for perception rotates into d for description, the former locked in the latter.

Where McCarthy relies on visual analogies, shape grammars embrace pictures and the Rorschach test, and thereby include McCarthy—and Minsky, as well. Shapes aren't visual analogies, at least none before calculating starts, and then one and sundry in retrospect, for as long as calculating goes on—description follows discrimination/perception and changes with it. Seeing is a far cry from AI and parametric definition in BIM. And in fact, this holds likewise today for computers and AI, with endless data, machine learning and training sets, and all the rest of it, that couch definitions in statistics rather than words, to diminish McCarthy's kind of constructive/parametric descriptions.) To get shape grammars, it was easy to copy generative grammars for natural languages like English or Greek—Noam Chomsky's pioneering work in linguistics—but it was evident right away that more than symbols (words) and recursion were required for visual calculating. In a generative grammar, "A" is the symbol A—although not like a rose is a rose, said out loud—and this is independent of the rules I try. And if I add two A's side by side, maybe they touch right leg to left leg, I still have two A's with nothing more possible—they're like the members of a set. And in fact, this is how Chomsky describes language today in his minimalist program that's framed in terms of internal and external "merge," with a "no tampering" clause to ensure that things don't alter when they go together.[26] Syntax is a set-like structure or tree in which words combine independently; they never transfer properties, or fuse and re-divide. Words are symbols and aren't like shapes—it's the same for spatial relations. In a shape grammar, "A" is the shape A—and also a triangle on top of a little table, with other parts of all sorts, maybe an artist's easel and in back, a handy shelf that opens to store paints and brushes, or a "Little Giant" two-step ladder that's being opened/closed, an A-frame with a ground floor and a second story above, upside down, an antelope's/bull's head, and anything else that I see and choose to pick out. There are indefinitely many different parts depending on the rules I try, even indefinitely many A's with shorter and shorter legs, until only the triangle is left. (Are A's like McCarthy's lines? What about the two squares in the rule in A1, when I rotate the inside square to increase its size, and how do the K's and k's in the original squares work?) And if I put two A's together leg to leg, I can find more parts. There may be alphabetical ones, an M or an N, a couple of rows of jack-o'-lantern teeth (two triangles that rest on half-hexagons), a raven hunting scattered seeds (M with a central beak in the middle of two splayed legs), twin mountain peaks (M cut bilaterally, in accord with an inverted b for p) with a snow line in each, a baby finch's gaping beak (M undivided), cat's ears (too obvious), and anything else my eye can find or my finger can trace—simply drawing in an alternative way. This also works using big K's and little k's—in four K's with perpendicular spines and overlapping arms or any other limbs you fancy, there are a pair of squares, four triangles, other polygons with axial symmetry, and then a myriad-times etc. in different lines and segments; and in eight k's arranged in a similar way, the possibilities are exactly the same. Moreover, shapes in rules aren't any different than shapes given (drawn) separately. How about the rule

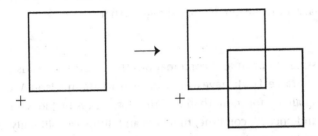

It looks easy enough, but does it add a square to a square, so that they intersect, both the same size on a common diagonal

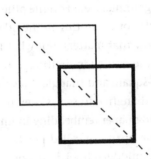

or replace a square with two L's

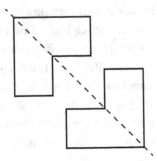

or these alternatives in parallel? Is there any end to this? And what about the left-hand side of the rule—is the square always a square, or something different? What's going on? Well, whatever parts I choose to see, in terms of the rules I actually try. Embedding is the reason why this works. In a generative grammar, embedding is strictly "identity"—A's are units, Chomsky's "atoms of computation" or "word-like elements" that are necessary for recursion. No one disagrees that

> an A is an A

no more or less; alone or in any combination, the identity of A's is never in doubt, they're always the same. I only have to look once to decide an A is an A. In a shape grammar, embedding allows for more than identity does. An A includes any part I can see (embed) in it, and once A's combine, they fuse and disappear—literally, they lose their identity, so that I can see other things, even A's. I can look as many times as I please and never be sure an A is an A, or how many A's there are. Neatly, identity is a special case of embedding—shape grammars allow for both relations, limiting embedding to get identity. This separates visual calculating on the one hand, and symbolic calculating (Turing machines, generative grammars, etc.) on the other hand—open-ended ambiguity in the one is simply monotonous counting in the other. With embedding, rules do what I see; it's what I value now that matters, not what I've drawn before—there's no memory of that in shapes. Those trained in drafting in the old-fashioned way with HB pencils or pen and ink, and T-square and triangles grasp this perfectly—they draw the longest lines they can find and are free to see as they please. It's uncanny how shape grammars come down to tracing paper (embedding in lines) and basic drafting (fusing lines together)—drafting is one of the so-called practical arts that boys like me were required to learn when I was in middle school. Many of us were destined for vocational education and hands-on jobs in the trades, instead of higher education in college for rocket scientists and ingenious engineers. I often wonder who got shortchanged. Learning to draft didn't harm me in any way—in fact, it was good fun seeing how it worked. Is embedding/fusing any less intuitive than counting or less rigorous and secure, as the basis for calculating? Can aesthetics replace logic? Don't be silly, learning to use logic in coherent arguments is the implicit standard for critical thinking in core subjects in college, and for lots of things later in life. Logic is the signal hallmark of an educated person. Is aesthetics ever a prerequisite for anything that's taught? (Ivan Sutherland rejects drawing with pencil and paper outright in "Sketchpad"—his groundbreaking invention is the prototype for all of the computer-aided design or CAD in use today, including trendy parametric modeling tools and BIM.[27] Sketchpad relies entirely on computer structure, with visual analogies or combinations of units, as stored descriptions of shapes for recognition and use, in order to tell what shapes are—McCarthy's slogan fits the bill to a T. In stark contrast to this, "Pen and ink or pencil and paper have no inherent structure. They only make dirty marks on paper." Pencil drawings in ordinary drafting are only lines, etc., and wait to be described—they're outside of McCarthy's parametric sweep. The gap between Sketchpad and pencil drawings can be measured in another way, too, in terms of figuration in Barfield's original and final participation. Sketchpad relies on computer renditions of idols—collective representations in final participation that are settled once and for all with no incentive to change. Pencil drawings, however, have only figurative sense and value—they mean what they do personally in original participation. The vast ambiguity in this is an impossible problem for computers to solve. It's the same for the Rorschach test, and the reason why von Neumann's visual analogies are doomed from the start, and fail. Nonetheless, Herbert Simon, Minsky's and McCarthy's compeer in AI, embraces Sutherland fully, in *The Sciences of the Artificial*. First, Simon praises Sketchpad for allowing intentions and goals to be fixed via constraints, to which drawings and their parts must conform—certain that every drawing entails a hierarchy of parts for "observation and understanding," in his own version of McCarthy's slogan. And then, Simon tries something unexpected; he concludes that design is like oil painting, with spots of pigment applied on board or canvas in a constantly changing pattern, so that intentions and goals are forever in flux. Simon is sincere about this kind of creativity, with no hint of absurdity, incoherence, irony, or paradox—Sketchpad, and drawing and painting are "combinatoric play" with building blocks or units of many predefined and richly varied kinds.[28] I guess this works if there's a unit for everything, but if this is so, what happens to calculating? And how does anyone count everything, or make any sense of it? No doubt, von Neumann would scratch his head and laugh. The whole point of drawing and painting is that shapes, pictures and the Rorschach test, aren't visual analogies—there are no units for shapes, without bringing an abrupt end to seeing. Combinatorial play, whether in Sketchpad or not, neglects what matters the most in art and design, because it takes away the looking—the unknown and nameless parts for tricks in fantasy, make-believe, pretend, and reverie at the quick of original participation, and the illicit impulse for the occult, pantheistic, and spiritual that just isn't calculable.) Shapes alter not once or once in a while, but for every rule I try—and this includes every identity $x \to x$, as well. At the risk of seeming oddly romantic, seeing is alive as heartbeat and breath in the in-out pulse of an "embed-fuse cycle" for new perception. It's seeing then doing, repeatedly starting over, fresh and entirely brand-new—

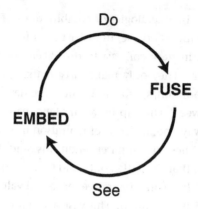

When I try a rule, no matter which one, the part I see is for the first time—in a cycle
that's embed-fuse, embed-fuse, etc. Visual calculating in shape grammars unfolds in a
metabolic process; it isn't mechanical repetition with fixed components and standard
parts, or recursion with atoms and words. That's the beauty of calculating with shapes
and not symbols, to go beyond units and calculating as it's supposed to be—without
giving up recursion. This lets art and design all the way in, so that seeing (drawing,
painting, etc.) overtakes thinking as the way to figure things out.[29] (MIT's motto is
*Mens et Manus*—the English translation of the Latin is Mind and Hand. This ties the
abstract and the concrete together, or theory and practice. It's the first thing to learn as
an incoming student, to solve problems and to think in "the MIT way." It's also how
MIT engineers invented CAD, with visual analogies—Steven Coons's "patch" to stitch
surfaces together; Douglas Ross's "plex" to define things in terms of data or indivisible
units combined in structures for algorithms; and later, the derivative work of Sutherland on Sketchpad, and of others on solid modeling, etc.[30] Actually, Coons has a nice
sense of engineering design that starts with a "graphical description" in which "the
designer's understanding of his creation is almost visceral instead of intellectual . . .
neither a well-turned English word description, nor a . . . mathematical formula . . . he
'feels' his creation" as it unfolds in sketches and drawings. But from the get-go, Coons
limits sketches and drawings in "building blocks of structure"—initially in "some
nebulous assembly" that cyclically comes together in a precise configuration.[31] I guess
structure is all he can think of, in the way designers, architects, and engineers, etc.,
automatically do with computers. And structure is key everywhere nowadays—in CAD,
computers manipulate descriptions and representations for practical ends, efficient
making in manufacturing and assembly, and also maintenance and repair. To architects and engineers at MIT and not, CAD and BIM are "data organization and operators

suited to that organization."[32] It's an inspiring vision for a perfectly managed world. Shape grammars try another method more attuned to what the eye feels, in impulse and intuition that's "visceral instead of intellectual"—shapes are changed on the fly in untold ways, to eschew the visual analogies and invariant structures that count when it comes to computers, and everything they're for. Nonetheless, shape grammars with parallel description rules provide the "refined details" of working designs.[33] Instead of Mind and Hand, shape grammars need something else to go on, to affirm visual calculating and to highlight it. The artist's formula Eye and Hand is a pretty good bet, to ground practice in aesthetics and new perception that skirt theory in description and representation. The resulting analogy is evident at once, and worth special notice—cancelling out Hand as a common factor, it's

CAD : Mind :: shape grammars : Eye

Then rearranging the four terms to align with my original analogy for shapes and symbols in the preface, there's the transposition

shape grammars : CAD :: Eye : Mind

in a quick summary of everything I've been saying, where CAD is BIM and AI, too. The mind settles on descriptions that the eye intuitively ignores, to see in extravagant ways—proof once again that seeing supersedes thinking, as shapes supersede symbols. The three-tuple Eye, Mind, and Hand, and its permutations are equal and inclusive, and no doubt, they're prudent, but they obscure the real value of switching from symbolic calculating to visual calculating. Somehow, Mind takes over, convinced that its descriptions are key—in definitions, graphs, and hierarchies or trees that come with a lifetime guarantee to discourage tampering. Mind, in words and language, drives seeing—that's what McCarthy and Simon urge for computers—with no way to reverse this to go from seeing to words, as the artist's formula allows, neatly in description rules and in other reciprocal processes, maybe with topologies. But in fact, Eye and Hand hold more than anything Mind and computers achieve. What computers do is for free in shape grammars—symbols and codes are a special case of shapes. The MIT way brims with confidence, and is widely and rightly admired, no matter that it's vexingly incomplete, riddled throughout with blind spots—misleading and somehow lacking all visual grasp, probably just another hat. It needs filling in, lest it miss what's easy to see.)

Minsky, McCarthy, Chomsky, and many others helped to get shape grammars off to a good start, with their enthusiasm for recursion and calculating, but with their techniques and biases rooted so firmly in logic and symbols, they missed embedding. The oversight was a dead end for art and design—their path wasn't the path forward. Other influences, however, came later, retrospectively after shape grammars were invented—in particular, William James (the marvelous discussion of sagacity and embedding in *The Principles of Psychology* that I extend to shape grammars in *Shape: Talking about Seeing and Doing*) and the wonderful Oscar Wilde.[34] I have a fantastic Phil May drawing on my office wall; it shows James McNeill Whistler replying mordantly to Wilde—who's amused, smoking a cigarette, how cool. I like to think that Wilde and Whistler would see immediately what shape grammars are all about. In the "Critic as Artist" and elsewhere in *Intentions*, Wilde talks effortlessly about poetry and visual art, and the critic as artist, as if he were describing visual calculating in shape grammars.[35] Here was an aesthetics that traced my route from symbols to shapes—eager for change. In fact, Wilde and von Neumann overlap, with uncanny fidelity. It seems that pictures and the Rorschach test are alike, exact copies almost word for word—different in two, nine letters and four, 13 in all. There's some crossword-like fun in this—"The one characteristic of [any] _ _ _ _ _ _ _ _ _  _ _ _ _ is that one can put into it whatever one wishes, and see in it whatever one chooses to see."[36] Is this Oscar or Johnny, Wilde or von Neumann? Is it art, or calculating? It must be a "beautiful form." I guess visual analogies fail in any aesthetic/recursive rule for a beautiful form or the Rorschach test—and the trick goes on and on, the same for any arbitrary blot, elaborate fret, entangled line, haphazard mark, intricate maze, irregular part, luxurious plan, segmented tape, or undivided unit. This is probably awkward in places—maybe the idea is silly—but my list is alphabetical for both words in each pair, with 121 combinations in all when I add in beautiful form and Rorschach test, and 156 with aesthetic/recursive rule in the mix. That 13 blank spaces can be filled in, in so many meaningful ways in a single sentence is impressive ("one can put into it whatever one wishes")—and the game continues, for any ambiguous film, play, poem, or song, important book, list, text, or word, or memorable epic, saga, tale, or yarn. Nonetheless, words (symbols) pale in comparison to what's possible with rules and embedding in shapes. Shape grammars unlock Wilde's "critical spirit" to unite the critic and artist—his aesthetic formula for this is to see things as in themselves they really are not.[37] It goes for every beautiful form, and for the Rorschach test, as well, anticipating von Neumann. Pictures are definite (finite) things that can be described in indefinitely many (infinite) ways, and plays and poems, etc. run the same course. Shape grammars exploit the recurrent ambiguity in seeing and new perception—forever vast and open-ended. There's a "moment-to-moment flexibility in the treatment [definition] of facts [parts]"—both in day-to-day politics (the line is George Orwell's in *Nineteen Eighty-Four*), where sense is easy to distort with conviction (sincerity) and not, and also fully in art and design. Isn't it strange how what matters (truth and beauty) always seems ambiguous? Visual calculating in shape grammars is an effective way to get used to how this works. The ongoing risk of "chaos and inanity" may be unsettling to many and even scary and terrifying. But insisting on an administrative standard, persuading (educating and training) all and sundry to see correctly in one way with rote finality—as some in America are encouraged to speak correctly—is not an alternative, merely dispensing privilege and safeguarding false security. Keeping to my addition to Pound's formula, art/calculating is charged with meaning, as rules change what I see with wanton vitality. The equivalence between art and calculating is a splendid example of Wilde's critical spirit at work. In whichever way the ratio is defined, each is seen as in itself it really is not. Then again, this also makes it possible to deny shape grammars totally—the computer scientist does because they're so different, unexpected and impractical, strictly for art and design, and the artist/critic/designer follows suit because they're only for calculating and so must be empty.[38] It's a riddle why people with clashing points of view find it so easy to agree—shape grammars are useless. That's a big snag for hats—what's interesting goes entirely unnoticed in their differences, with no awareness of missing anything. This may explain why there are barely a few who grasp what shape grammars actually do in the embed-fuse cycle, who make use of them eagerly as the way into art and design, and who exploit them as a way to calculate that doesn't rely on counting. There's plenty in shape grammars for the computer scientist and the artist/critic/designer alike—when they're taken seriously, and when they're not. Maybe that's the real test for calculating, that it goes on, no matter what.

So, who are shape grammars for? I started out using them to paint—I remember warming up on Paul Klee's drawings in *The Thinking Eye*, notwithstanding the synonymous formula Eye and Mind that puts limits on shapes in terms of visual analogies and other symbolic descriptions of their parts (McCarthy's slogan once again and equivalently, Chomsky's merge and no tampering, both for Mind expressed fully in words and language)—and then for visual design, mostly in architecture, to calculate anything from ornamental designs (for example, irregular Chinese ice-rays and symmetrical window lattices) to Palladian villa plans replete with articulation (porticos and wall inflections), windows and doors in enfilades, and decoration, and even to present-day projects like Alvaro Siza's *Quinta da Malagueira*. But shape grammars are really for anyone who wants to design visually or to study how this might be done, whether it's in art, architecture, graphic design, urban design, or engineering and product design, or best in a discipline of its own with a nifty mathematical slant. Shape grammars work with description rules and weights to link form, function (use), material (color, etc.), and alternative ways of making (from assembly and manufacturing processes, to painting and drawing) in multiple relationships for any size, at the human scale, and the nano and humungous. This goes beyond computer modeling as usual in CAD, BIM, and AI—there's no loss of anything that's visual along the way. Moreover, it serves to divide shapes on the fly, and to record changing divisions in terms of retrospective graphs, hierarchies, topologies, and kindred descriptions, representations, and structures. Shape grammars trace what I see in different ways, and let me overstep any discontinuities in analysis and synthesis; they describe things already known anew, as in themselves they really are not, and they make new things of many and varied kinds in this process and also from scratch. (To the poet and critic Samuel Taylor Coleridge, synthesis implies "indifference" in organic unity, that G. E. Moore doubts in the preface.[39] Every shape is the synthesis of a thesis—two squares—and an antithesis—four triangles—but the shape/synthesis doesn't care about these opposing descriptions or visual analogies. There are "myriad myriads" like them primed for instant use. All of these save the appearances. Better yet, this is "the effect of reducing" multiple descriptions to a single shape, and it's the same for "reduction rules" in shape grammars, to embed and fuse, for example, in the left- and right-hand sides of the identities I talked about in A2.[40] Coleridge ascribes indifference to things that are alive and that grow, to plants and poems, and possibly to you and definitely to me. Indifference fits shape grammars perfectly, although not every generative process. At one time, I was 99.44% sure that all generative grammars had lives of their own; thanks to Coleridge, I'm now more prudent about what organic/romantic analogues hold—they're fine for pictures and poems, but they're senseless in symbolic calculating that's outside of the embed-fuse cycle for shape grammars. My original analogy in the preface, that connects Coleridge to shapes and symbols, entails this and more. Analogues and metaphors may help to explain shape grammars—to tie them to familiar things and not—but shape grammars are indifferent to this, ready to assimilate descriptions that vary arbitrarily and clash. Maybe that's the reason why shape grammars are different than hats—hats aren't indifferent. There's no single right approach to shape grammars, but playing around with them on their own and trying not to anticipate too much what they actually do is a productive place to start—to build new intuitions that run strong and sweep wide.[41] This isn't so easy; I still find myself ensnared by old ones. Relationships that I expect to be this way invariably turn out to be that way, and then yet another way. With shape grammars, it's only routine to be surprised—here and now, and with hindsight looking back.)

**Q4.**

How has research into shape grammars evolved since?

**A4.**

Shape grammar research has taken off in many directions. There's a lot of work—some of my own but mostly of others—that uses shape grammars to investigate style and typology in art and architecture—what's a Palladian villa, what's a courthouse, etc.?—and for forgery, that makes it possible to franchise different architects and artists, and to put money in the bank. Somehow, I'm sure I'd rather keep the beautiful forms. Shape grammars have also been used creatively both in architecture and in product design—their promise in practice is immense, waiting for designers to risk it and try them out, boldly and confidently. There's a striking body of work, too, that shows how visual calculating in shape grammars can be used to focus art and design education. Some of this depends on recent theoretical results that define schemas as heuristics for rules of varied kinds and establish their relationships. The rudiments are too tempting to skip. I promise—the details are painless in a short diversion that's scenic, and worth the extra time and a few words. In fact, one of the examples I like to use for schemas prefigures what I want to say about how insight and imagination work, and how they're assimilated in full force and no decrease in shape grammars. A schema

$$x \rightarrow y$$

has variables, here x and y, that are assigned shapes as values to define a set of rules. Primary schemas for parts, transformations, and boundaries imply others, in subsets, copies, inverses, adding, composition, and Boolean expressions.[42] For example, adding the schema $x \rightarrow x$ for identities that's a subset of the schema $x \rightarrow prt(x)$ for parts, when $prt(x)$ is x, and the inverse of the schema $x \rightarrow b(x)$ for boundaries explains the artful workings of the coloring book schema

$$x \rightarrow x + b^{-1}(x)$$

that fills in areas and saves their outlines—for alternate values of the variable x in a building plan, there are walls or poché, or rooms and spaces. For this shape

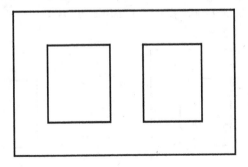

the two rules in $x \rightarrow x + b^{-1}(x)$ are these

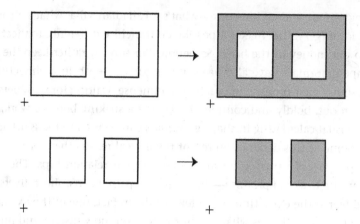

to fill in areas with grey. This gives

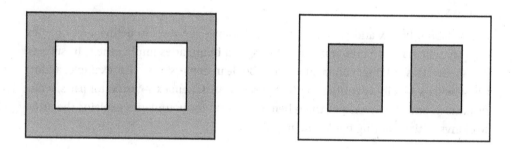

(In oddly artsy/geeky moments—sometimes, I simply can't help myself—I put the composition $b^{-1}(prt(x))$ in place of $b^{-1}(x)$ to get the alternate schema $x \rightarrow x + b^{-1}(prt(x))$, because there's a single value for x, namely, the shape itself for coloring in. The new schema switches walls and rooms for two distinct parts

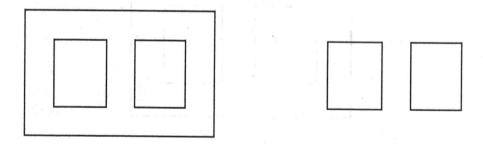

and there are five other parts for which the inverse boundary schema is defined. Three of these cover everything involved—

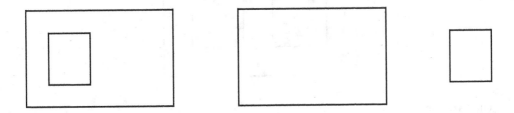

The coloring book schema is a subset of $x \rightarrow x + b^{-1}(\text{prt}(x))$; both schemas go nicely with weights, and are similar versions of the summation schema I try in the next parentheses. And in this way, $x \rightarrow x + b^{-1}(\text{prt}(x))$ is a subset of the inverse $x \rightarrow \text{prt}^{-1}(x)$ of the part schema $x \rightarrow \text{prt}(x)$. The rectilinear boundary

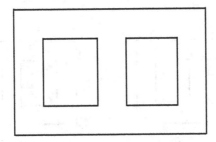

is part of the shape

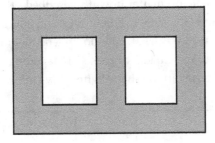

and also part of the shape

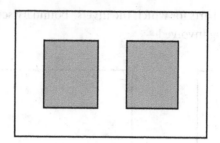

The two rules

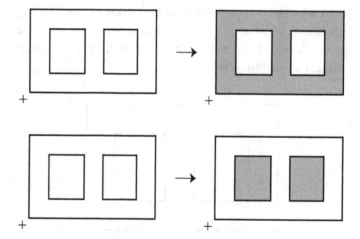

can be described in alternative ways. It's not uncommon for a rule to be in separate schemas that define rules distinctly—a rule is in x → prt$^{-1}$(x) anytime its left-hand side is part of its right-hand side, no matter how it's described otherwise. Words are good heuristics most of the time, but they can be misleading and not always the best way to think about rules; what I see is what rules do, independent of anything I say about them.) I can use weights to vary shades of grey, so that adding the grey in a rule to any grey in a shape makes the grey in the shape lighter. My twin rules and this one

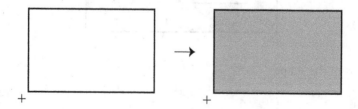

also in the coloring book schema, define indefinitely many shapes, including these two

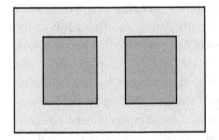 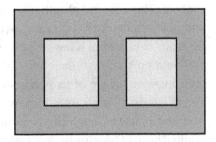

I used this idea when I started out painting with shape grammars and generative specifications 50 years ago (see note 29). This is an electronic version of one of my pictures from 1968—

Actual pigments on canvas are better to see and do, but this isn't something for a book with multiple copies. (There are lots of other possibilities, as well. In particular, r is part of n—not as a letter or symbol but as a shape—so the alphabetic rule n → r is in x → prt(x), ready to try in Pound's *ABC of Reading*. A number of years ago, I used the rule h → r to show that cheating includes creating, as changed includes charged. This changed cheating and charged it with new meaning to empower artist and novice alike with a creative method that actually works, and that's fun to try. Everyone copies, even if few will ever say so—it's just not a practice that's encouraged. You do it on your own in secret and keep it to yourself, first when you're stuck and then to go on, aware of its potential and confident of its value. Copying—cheating and plagiarism—is an unfailing way to delightful things.[43] It's key in both education and practice; inventing is seeing and copying. Leon Battista Alberti fixes the original locus of copying in *On Sculpture*, when artists, "diligently observing [seeing] and studying," were led to find in nature, "in a tree-trunk or clod of earth," outlines to trace and complete in drawings, pictures, sculpture, etc.—by adding to, taking away, or otherwise supplying what seemed lacking.[44] What would Simon say, and Sutherland?—so much for hierarchies and computer structure that discourage tracing outlines to suit yourself, to add to and take away. This kind of creative copying—maybe all copying is creative—extends neatly throughout art and design from tree-trunks and the like to pictures finished or not, mine or someone else's, to whatever I choose to see in a beautiful form or in the Rorschach test. In fact, copying may be an important source of personal style—when artists and designers copy themselves. Copying is straightforward in visual calculating in shape grammars—it's plainly there from the get-go in the embed-fuse cycle, at the quick of how rules work. And copying is key in rules themselves, for example, when an identity x → x and the primary schema for transformations x → t(x) combine in the addition schema x → x + t(x), to add a shape x and a copy of it t(x). In schemas, rules can be described in sums of changes to shapes and their parts—I'm free to copy in this way and in that without any fuss, for something that's fresh and new. But whether you buy into copying fully, as a legitimate method for productive work in art and design or not, isn't really that important— schemas do what they do indifferent to rumor and personal opinion. They are what they are, no matter how they're described—and there are myriad ways to do this. Let's see what the summation schema

$$x → \Sigma \, F(prt(x))$$

does to a given shape C, where the variable x is C or anything in it, and F tells which rules apply to parts of x, say, rules in the schema for transformations or rules in the coloring book schema. All of x is taken away, as its parts—any of them, even more than once—are replaced with new shapes for rules in the schema prt(x) → F(prt(x)). Every rule A → B is easy to define in the summation schema—start with the shapes A and B, then prt(x) = x = A, and F(A) = B. And this inclusiveness isn't empty. I can use the schema to describe A → B in any way I like, and keep an open mind; the rule is the same in indefinitely many sums, for rules in the schema prt(A) → prt(B) that divides A → B into parts—besides A → B, maybe rules in A → prt(B) or prt(A) → B. Shapes/parts, different and not, are picked out and changed in a single try with multiple rules, in a process that seems parallel at first and then sequential. The descriptive intent for each of the terms F(prt(x)) in my original schema highlights a separate rule, until the shapes in the terms fuse/sum in a seamless result—it's like this when visual analogies and spatial relations are given to describe shapes. Then A → B is pictorial, like every other rule in the comprehensive schema

   drawing1 → drawing2

from A1, for any pair of shapes. Identities in x → x are telling; they parse/divide shapes in graphs and trees. Such diagrams turn unstructured shapes with indefinite parts into structured sets with definite members. And identities meld as one—I can keep track of those I use in the terms F(prt(x)) and how they're tried to trace the wiles of perception, and then forget/omit this adding them up in a rule. Copies of the identity

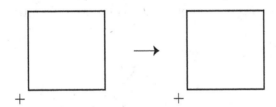

for squares divide the shape

all at once in a tree for two squares, one inscribed in the other—

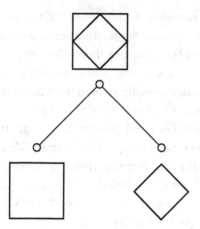

And when I do copies of the identity

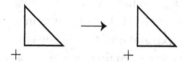

to resolve triangles all at once, I get an alternative tree that distinguishes four corner areas—

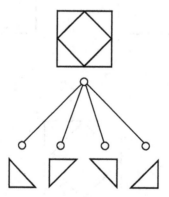

The vertices in these contrasting trees differ in how many and what—two is two less than four, and squares aren't triangles. This blatant inconsistency, Coleridge's thesis and antithesis, is merely one of many clashes that measure half-knowledge and the reach of negative capability in x → Σ F(prt(x)); but sure enough, the incon-sistency disappears in a synthesis—for both trees, the schema defines the same identity

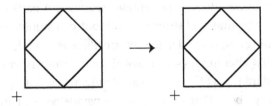

that doesn't divide the shape

Shapes are shapes—alone and in rules, they're indifferent to anything I say they are. Nonetheless, there's always more to see. I'm free to pick out parts at will, never affect-ing what's there, flipping back and forth in the embed-fuse cycle, in unsettled percep-tion. Alternative versions of the identity

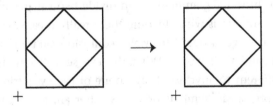

include polygons, letters, and other parts. I can have my cake and eat it, too, taste by taste—in fact, shapes can be reconfigured in different sums of identities for function, purpose, and use, changed yet unchanged in description rules. The use of identities is one of the best ways that I know to show why symbols—invariant definitions couched in visual analogies—fall short whenever I calculate with fickle shapes. The schema x → Σ F(prt(x)) helps to classify rules in heuristic subsets, as it combines (adds) rules for confluent use—every rule has myriad composites in this way. They inform seeing to relate designs that look poles apart in similar processes, as the values of x and their parts, and transformations vary—there's proof in Exhibit 3. Rules in the schema are global and local all at once; they go for large features and small details, in open-ended improvisation. In the pure flux of perception, shapes and parts are modified mutually, beyond constant syntax and structure, and routine combinatorics—changed and charged with meaning and value. There's the artist's/poet's reordering of experience in shapes and parts that alter in sync—in magical "transmemberment."[45] In the schema, rules conflate standard types of symbolic calculating, when shapes fuse to blur distinctions for units and symbols. This already goes for parallel vs sequential calculating and equally, for top-down vs bottom-up—and when counting is involved, it extends to context free vs context sensitive, and the certainties of complexity. Use of the schema exploits the inconstant relationship between shapes and symbols—separate processes are distinguished in the terms $F(prt(x))$ that aren't differentiated in the rules the schema defines, or in the shapes these rules produce as they're tried. The schema describes rules to make shapes that needn't be described in this way. There's no history or other restrictions on what I can see next; only rules will tell, independent of anything I think I've done. I'm constantly starting over—distinctions in the schema are lost completely as soon as shapes fuse, so that there's nothing to block my way. Prior distinctions may be memorable and also perfectly true, and a lot to talk about, but so what—it's OK to change my mind, at any time I wish and as I please. That's why embedding is key and why shapes fuse, and why rules are the way they are in the embed-fuse cycle. It's also worth trying the schema $x \rightarrow \Sigma\, F(prt(x))$ for rules that add shapes rather than divide them, to see how this unfolds—although dividing and adding aren't as far apart as they may seem. In shape grammars, I can divide two squares by adding them, or maybe adding two squares is what dividing them really means. Didn't I just do this or something like it to get a tree? This weird confusion isn't any reason to stop; it's merely fooling around with words. What the eye sees makes the only difference that matters, indifferent to anything I say—none of the eye's tricks is captured in a single formula expressed in symbols or words. After all, all of the eye's tricks play out in every detail in terms of the rules in $x \rightarrow \Sigma\, F(prt(x))$ that delimit perception without limiting what comes next. Examples show the schema's heuristic power and sweep, in drawings that invariably defy the words I try on them. There's surprise and sometimes, the embarrassment of not having seen it. I like to tell myself that my discomfort is misplaced; rules in the embed-fuse cycle supersede anything I can ever say—it's something you get used to, and eventually learn to like. Let's suppose that x is the square and prt(x) is x four times, then for F given by copies of the addition rule

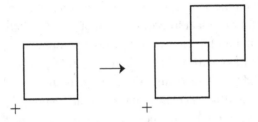

in the schema x → x + t(x) that combines an identity and a transformation—here, a translation on the diagonal of a square—I get the shape

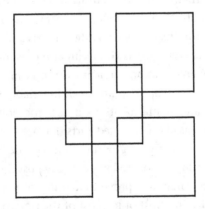

How many squares are there now? What's next, doing this again for the four small squares that aren't described in the addition schema I used to define F? How about for the extra big squares, or for different combinations of big and small squares? There are many ways to see this, with squares and without. And what happens if I decide to include one more rule

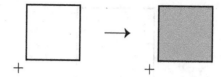

in the coloring book schema, and grey gets lighter as it's tried? And this is just warming up. Still, enough is enough. Technical agility isn't the goal right now—that's left for later in Exhibit 2—rather it's for something intuitive and less math-like.) Whenever they're used, shape grammars incorporate insight and imagination in visual calculating. Sometimes, it's helpful to look at insight in terms of gestalt effects, for example, figure and ground reversal in building plans, when walls and rooms switch back and forth, in ice-rays and other Chinese lattice designs, and in painting and sculpture—maybe Ellsworth Kelly and David Smith. Of course, there's more than a single gestalt switch when shape grammars are used to calculate—shapes are ambiguous through and through. (Marjorie Garber urges the same for poems and language broadly, in her review of George Lakoff's and Mark Turner's "mappings" for metaphor.[46] To Garber, metaphor is "a first-order, not a second-order, phenomenon . . . not simply a clever kind of code [mapping] . . . enabling the nimble linking and blending of commonly held thoughts." Like shapes, "words and rhetorical forms are themselves unstable, [with] alternative and often antithetical narratives of their own." Garber turns to Gestalt psychology and figure-ground reversal for this instead of Coleridge's thesis and antithesis—to put metaphor beyond mappings and code, just as von Neumann puts the Rorschach test beyond calculating with visual analogies in an implicit turn to beautiful form. In fact, spatial relations for shapes show why mappings and such invariably fail. Encoding shapes to calculate with mappings and spatial relations misses the point—there's more to see and more that alters than this allows in fixed parts that are known in advance without showing how. The psychology of figure-ground reversal or the Rorschach test, however, doesn't seem necessary for metaphor. Today's cognitive psychology and cognitive science are mostly irrelevant. They may prove that embedding is inborn and as natural as using language—effortless to most of us most of the time, and usually unnoticed—and discover how the brain is able to do this wonderful trick. But regardless of what they're able to show, shapes are ambiguous. As long as shapes lack the cognitive/conceptual structure in visual analogies, there's little to say about what any of us is likely to see next, or so it seems to me. This kind of seeing is "first-order"—there's nothing in shapes to ground it. Isn't the creative use of language—metaphor and not, in ongoing experience—a mystery for similar reasons?[47] No doubt, shapes and metaphor overlap in key ways. Wilde sees no future for psychology and science in our world; they can't explain "well-dressed women" (architecture) or "grapple with the irrational" in this—perception has no bounds. Wilde prefers music, and pictures and poems, asking only for a beautiful form. "Beauty is the symbol of symbols"—

> Beauty has as many meanings as a man has moods. Beauty is the symbol of symbols. Beauty reveals everything, because it expresses nothing. When it shows itself, it shows us the whole fiery-coloured world.[48]

For Garber, the Rubin vase is the "figure for *figure*," that is to say, for metaphor in poems, etc. Every metaphor is a beautiful form or the Rorschach test. Reading and rereading take insight—"All language is figure, and figuration: it is the idea that we can see through language [in rich and changeable ways] to encounter the real." And what's real in figure and metaphor hinges on the fickle results of perception and on personal method, on ambiguity and the inconstant arc of half-knowledge in negative capability. This isn't mappings and code, either in minds or computers. In both, the real has a constant structure that's stored in memory; once meaning for pictures and poems is resolved, it isn't expected to change—but then what? "Literature is figure," and figuratively, figures are shapes. In shape grammars, shapes have no structure in themselves—they alter with the rules I try, in accord with my personality and mood. Meaning is a rich, ongoing, open-ended process.) I bank on embedding for this, and for Alberti's way of seeing "in a tree-trunk or clod of earth" (Rorschach test) and what it implies—as Shakespeare's Theseus fears, "a bush supposed a bear"; to Alice's surprise, a Cheshire cat that appears and disappears at will (whose?), as a single all and a plural not all in the tangled branches of a tree and their interstices; and of course, in Barfield's chaotic and inane world, a dog with spare limbs fast at play in a vegetable marrow. (I'm always amazed that our ancestors managed to transform the shifting shapes in dense forests and wild woods into vast and sprawling cities and urban spaces, and that we're able to cope with this relentless ambiguity.) Shakespeare's "strong imagination" conjures "such tricks," in pictures and poems—in fact, everywhere in art and design. Picky questions of plagiarism and fussy exegesis aside, this strikes me as the quick of Coleridge's extraordinary "esemplastic power"—"the *mode* of its operation" is to "shape into one" for new perception, as in the embed-fuse cycle in shape grammars. (Being a poet and critic/artist, Coleridge is expected to copy, and to steal and to miscon-strue shamefully—beauty is truth. Art isn't for the scholar; Coleridge is very much on his own in the way he frames imagination—greater than anyone before him, expansive in the embed-fuse cycle.) Esemplastic power (fusing) is the source of imagination and yields it fully—

[imagination] dissolves, diffuses, dissipates [boundaries, to fuse memoryless/structureless wholes], in order to re-create [embed] . . . It is essentially *vital*, even as all objects (*as* objects) are essentially fixed and dead.[49]

This puts objects (shapes and things) in motion (flux), as it activates seeing. Parts alter freely—they fuse and re-divide in sense experience, in no way ever fixed and dead. Such transitions are progressive, but with neither ends nor goals—there's the pulse of life. (Whether this ongoing process repeats known divisions or lets in new ones is mostly irrelevant—because parts fuse with no memory of their boundaries, the same kind of calculating is a must either way. Not everyone is easy with this. It would seem that Barfield reaffirms known divisions in final participation and regrets the extravagance of new divisions when participation is original. Other distinctions are out, as well. In particular, whether to fuse and divide is conscious or unconscious doesn't matter—there's no difference in calculating. In unguarded moments, though, I use the Cartesian product of the pairs known/new and conscious/unconscious as a rough-and-ready way into a taxonomy for Coleridge, Barfield, and many others. When my guard is down, I'm liable to see just about anything. And first impressions are effective more times than not—it's amazing how odd things can be in art and design. Figuration in varied forms is key—caricature, distortion, exaggeration, metaphor, etc. Imagination fills the soul! And with a little sobriety, simplifying to the limits of recognition is a rich way to unpack strange relationships. For example, neither Eastern thought nor Western sophistication seems particularly well disposed to imagination—it's mostly a tempting nuisance. Everything in the East is one, with no compensating emphasis on embedding or individuation, and in the West, atoms merely combine and never fuse. That's my gloss of it—each a distinct part of the embed-fuse cycle. But philosophies East and West merge and dissolve, coadunate in shape grammars—in the ongoing pulse of the embed-fuse cycle. The trick is to keep this process in motion, so that there's never a pause. It doesn't work to assume units just for the time being, to allow in logic and symbolic calculating in Turing machines. This invariably ends in obligatory truth and unpleasant results.) Without embedding and the insight it brings, it's hard to go on to anything new; without imagination, "that synthetic and magical power," the critical spirit lingers only to wither and die. The literary critic I. A. Richards traces Coleridge's imagination much as I do—"earlier acts of perception [that] . . . have come into being, been formed, by earlier acts of Imagination [are] . . . *re*formed . . . integrated, co-adunated into new perception."[50] In shape grammars, constituent parts (units) needn't limit what I see—I can embed two squares in four triangles once they fuse. And in exactly the same way for poems, *"Words [triangles] are not necessarily the units of meaning [squares]"*—poetic imagination tests language "above and beyond . . .

common use."[51] I know what I mean when I see what I say. For all of us, this is a real possibility that plays out freely—pictures are what we see (embed) in them now, and they change when we look again; poems mean what we read and talk about now, and they alter as we go on reading and see (perceive) in alternative ways. (Some prefer to say misreading, although this may be a mistake. Does anyone say mis-seeing? Suppose he/she did—what difference would it make? Isn't mis-seeing also seeing and so, believing?) New perception never ends; what we find in pictures and poems is neither everything nor final—that's why pictures and poems endure in imagination with its remarkable esemplastic power to re-create. Richards is full in, over-the-top keen on imagination, as it influences established beliefs and values. It makes a real difference—

> [N]either Coleridge's grounds for [imagination] nor his applications of it have as yet entered our general intellectual tradition. When they do, the order of our universes will have been changed.[52]

And so, charged with meaning. In Coleridge's esemplastic power, there's a kind of active "projection"—we make up new parts to embed in things that fuse. (Others talk about shapes/parts that emerge somehow. Many see this when two congruent squares intersect on a common diagonal, to form smaller squares in their fluid interstices—moving squares in this way is what McCarthy does with lines in spatial relations, but how do new squares emerge or other shapes like L's without embedding/fusing? AI forgoes imagination for mere description—oops, I meant to say BIM.) It's a nifty trick, but for "convenience," Richards "dispenses with [it]." Not "denying the validity of [Coleridge's] projective account," he turns instead to practical criticism and to that with which the critic is most concerned—a structural analysis of "differing types of interaction between parts of a total meaning."[53] In the terminology of math and computers, he settles for a graph, very likely directed with strongly connected components. Of course, graphs are never the source of imagination in pictures and poems, or in any results in art and design; they're merely an effective way to describe and measure aspects and properties that arise independently in some other way. However, the method is seductive—few can resist it—even if it requires a steady hand and often leads to ruin as an easy substitute for imagination. But this isn't a problem for Richards—he always aims for what's important and isn't known to miss. He values imagination highly—"Doubtless the ideal case of Imagination is rare."[54] I wonder why. Is there something about imagination in itself that makes this so, or is it simply what graphs and the lure of convenience imply? It's pretty hard to tell outside of shape grammars—they assimilate Coleridge indifferently, as they fill in the details of Richards's total meaning in a reciprocal process with description rules to put in parts and to fix their interactions, in graphs and retrospective topologies. In shape grammars, insight and imagination are unavoidable; they come in automatically as rules are tried. Imagination works recursively, rule after rule for all to see. It isn't so much that imagination is rare, but that we tend to suppress it, preferring truth (security) and old memories (nostalgia) rather than risk new perception. Imagination unfolds neatly in the embed-fuse cycle. I can show exactly where it works and how, and why anything less muffs its tricks, graphs and topologies included—although many distrust my endless run of examples that must be drawn and seen, and that sometimes defy words. (A few of these examples—three in greater detail—are shown in Exhibit 2.) Nearly everyone I know, computer scientist and not, believes that symbolic calculating—combining given units (atoms, bits, building blocks, objects, primitives, simples, etc.)—covers all the bases. My neighbors in Brookline tell me the same thing—the famous biologist a few houses down on the right embraces "BioBricks" for Lego-like design in synthetic biology, while the rigorous computer scientist on the next street to the left lectures me zealously on the atoms of computation and their necessity in recursion. This seems perfectly correct, when you're outflanked on two sides. Symbolic calculating is the final standard, steeped in structure and averse to any ambiguity; there's no room for question or doubt. Its paired loci span everything with complete certainty—in axiomatic logic, and in Bayesian statistics and machine learning. Appositely, axioms and data trace a single path. And both fall short of insight and imagination that are crucial to keep art and design creative—and really, insight and imagination for science and engineering, too. The Gestalt psychologist Wolfgang Köhler and the cyberneticist Warren McCulloch were among the first to consider this, around 1950. Insight was missing in Norbert Wiener's statistical version of cybernetics in which feedback is key.[55] And this goes equally for McCulloch's groundbreaking calculus of nervous activity (neural networks) in which you can calculate anything you can describe in words, or likewise, with numbers and data.[56] (Neural networks, in one guise or another, are key today in brain and cognitive science and in computer science, too. This goes for the brain with only neurons to resolve objects/things, and it's mimicked in AI.[57] Still, neural networks aren't shape grammars—they're not about sensory data and neurons, but focus on shapes/parts and rules in the embed-fuse cycle. Linguistics does the same, leapfrogging aural data and neurons to emphasize words and rules. And yes, mutual relationships abound for data/neurons, shapes, and words.) It's no surprise that McCulloch touts every bit of cybernetics and calculating (computers and AI) with unreserved enthusiasm and unflagging confidence. His essay "What Is a Number, that a Man May Know It, and a Man, that He May Know a Number?" seems curiously appealing, right down to the palindromic (cerebral/hemispheric) symmetry in its title. The sequence of terms goes from number to man to man to number—maybe to imply reflectively that man is number and number is man, and that there's an underlying equivalence between man and computers (counting). Nonetheless, McCulloch begins with an honest confession that impacts AI through and through—

> But the problem of insight, or intuition, or invention—call it what you will—we do not understand, although . . . that is the problem I should tackle.[58]

(In *Cybernetics: Or Control and Communication in the Animal and the Machine*, Wiener recites a nifty children's song about how God with his/her matchless power can count both the stars in heaven and the clouds in the sky, and knows when all of them are there, sure that none go missing. Wiener's translation of the original German runs so—

> Knowest thou how many stars stand in the blue tent of heaven? Knowest thou how many clouds pass over the whole world? The Lord God hath counted them, that not one of the whole great number be lacking.[59]

Wiener uses the song to split science into disciplines, contrasting ways of analysis in astronomy and meteorology—definite things like stars and their constellations are ready to count one by one in exact number and move with unerring regularity in logic, while indefinite things like clouds with fuzzy boundaries and random motions are only secure in data and statistics. But the song also defines a distinct choice in shape grammars. First there's identity—calculating with visual analogies or spatial relations, where counting is one by one for members of a set, or for fixed/unit objects such as points that add up in shapes, just like stars in constellations. And then there's embedding and things fuse—calculating with shapes that are all like clouds to see/find faces, and animals, castles, cities, islands and seas, and sundry images and figures à la Alberti and the Rorschach test. Then ambiguity reigns supreme, as seeing supersedes counting and statistics—they're unstable and break down when boundaries dissolve and re-form freely and without prior constraint. Incredibly, Wiener's song puts God, cybernetics, and shape grammars together; the three merge as one in a comprehensive verse—the song is a beautiful form. Its lines trace counting, statistics, and seeing.[60] To God, counting is all it takes to survey all there is in Coleridge's primary imagination. Clouds are as definite as the stars—everything that's in heaven and on earth can be tallied one by one in a perfect census and a single number. Lacking God's power to count clouds, cybernetics turns to data and statistics to track the locus of what isn't definite and fixed—this is standard everywhere today in computers and AI, to discover the hidden structure of the world, and to describe and represent it. And shape grammars dissolve and diffuse divisions to fuse known parts into one—seeing and embedding replace counting and identity to include insight and imagination, Coleridge's secondary poetic-kind, and the critical spirit in new perception. But taxonomies aren't always so neat and clear-cut.

Sometimes, three make an asymmetrical two—counting and statistics strive for certainty in prediction and control, while seeing's inherent ambiguity guarantees precisely the opposite. I wonder if there's a convenient way to follow the meaningful paths in this vast array of interactions. It seems like a pretty good time to try a Richards-like graph for a total meaning, if only to show how worthwhile this is, even for a little song. My graph has 26 vertices—one for each letter of the alphabet—and too many edges to count, although once shape grammars are added in, it's clear that calculating connects to more than God or science. I guess this spells the end of physics and any chance for a theory of everything. Fantastic, except the complete graph is a monster that won't do for eye and hand. With its scattered vertices and crisscrossing edges, paths are hard to untangle and often impossible to find. It takes a computer to map them out and tidy up. Maybe a simple subgraph merits a serious try. At the heart of my graph, there's a subgraph in which clouds (unanalyzed phenomena) connect three pairs in triangular cycles—(1) God and counting, (2) cybernetics and statistics, and (3) shape grammars and seeing. These pairs fit in a regular heptagon, with clouds at their common vertex. Of course, this isn't all—for shape grammars, there's seeing and embedding plus (4) counting and identity, and an extra edge in a triangular cycle. Putting the new edge in place, to connect shape grammars and counting, makes it tempting to play God, giving up insight and imagination for visual analogies together with computers and AI—invariably, this spells disaster, nothing but error upon error and final collapse. It makes better sense to enhance the symmetry of my heptagon in a triaxial pinwheel, so that clouds are at the center. Now I can see exactly how Wiener's song goes—with no irrelevant details, in the *CliffsNotes* version of my original graph. There are meaningful paths to trace through four equilateral cycles. Three of the cycles in the pinwheel move counterclockwise, as the asymmetrical fourth goes the opposite way around in tragic misdirection. Maybe there really is something to this that actually works. It feels entirely objective; everything seems just right. Do subgraphs set out the units and component parts of meaning that add up to a total meaning—"manufactured, brought together, and endued with vital warmth?" I guess it's about time to take a close look—

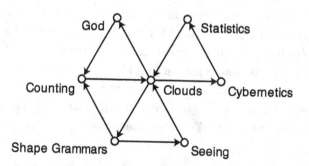

Graphs like this are OK without being monsters, as they highlight differing types of interaction—however, the method from song to graph to subgraph(s) is most likely incomplete. What good are graphs or trees or whatever structures I care to use, if I can change my mind about what I see at will, and want to factor this in? It's easy to put in vertices and edges for ambiguity and certainty—for things that go with what I already have—but how about other things that are outside of this, that seem incommensurable like the two squares and rectangle in the second example in Exhibit 2? In contrast, shape grammars are inclusive in the embed-fuse cycle—in fact, as a magical extension of feedback in cybernetics, in which sense experience is open-ended and unrestricted with nothing set in advance, in an expansive rather than secure process. Who would have ever thought that clouds held an alternative to God and Wiener's cybernetics in shape grammars, when there's embedding and shapes fuse? Isn't all of this proof positive for seeing and visual calculating, that there's more outside of counting and statistics, in open-ended ambiguity? In shape grammars, ambiguity is at the shifting frontier of extravagant delight, the Vitruvian aesthetic third that's above and beyond certainty and use. Calculating is for science in counting and statistics—and expansively in another way for art and design that entails science, as well.) Solving McCulloch's problem is a lot harder than it looks at first blush. Progress is elusive—maybe a shape isn't a number or data, or maybe insight isn't to be found anywhere in descriptions (visual analogies and ancillary structures). Seventy years later and still counting, McCulloch's problem has become a lasting constant in computer research and AI, older than Moore's law and well outside its scope. No doubt, there's always the hope of a big breakthrough, and breathless expectations. IBM's Watson is mindlessly hyped like this. Still, not everyone is impressed with its skill, even with its crushing win at Jeopardy; it beat the best of the human contestants, and both of them when their scores were added together—although not always first in this way. David Gondek, one of the leaders of the Watson research team, is quick to note that scant of Watson's mega power for calculating/thinking matters—

Insight remains handmade.[61]

It's exactly the same for Deep Blue, AlphaGo, AlphaFold, and machine (deep) learning programs of all kinds. Insight is ours to have, and to impart in retrospect—that vital switch in perception that opens up new ways to go on, independent of symbolic calculating and the vast resources of computers. (Aren't the rules in shape grammars ready for insight, and how to go on?) This is good news to many—computers aren't creative and aren't likely to be anytime soon, maybe not at all. Creativity is forever out of reach, as long as everything is kept only to a number or data. Rather than number and data, insight—intuition and invention—and imagination are needed to go on. The food columnist and blogger David Sax agrees; he tries his own formula in a recipe that blends in imagination for needed seasoning, as he stirs sensibly with eye and hand, unencumbered by the calculating/thinking mind in computer technology—

> Creativity and innovation are driven by imagination, and imagination withers when it is standardized, which is exactly what digital technology requires—codifying everything into 1s and 0s, within the accepted limits of software.[62]

But try to say more without shape grammars and the embed-fuse cycle, and Coleridge. (In the past, I've fooled around with C. S. Peirce's abduction as the instinct to imagine correct theories and hypotheses from scattered data, and a nice way to be original. It strikes me now that abduction aims for noticeably less than insight and imagination in shape grammars. With its focus on verified facts and units, abduction misses what's quick to see in art and design.) STEM subjects are a must just about everywhere in education, usually at the expense of the humanities, the arts, physical fitness, and other subjects that don't divide everything into units. But without insight and imagination in art and design—without shape grammars—STEM may seem meaningless, mere counting waiting for a reason why. It's textbook, rote, mechanical, and dull, with scant opportunity to see past visual analogies, to be creative, or to find out how. To use Coleridge's famous distinction, it's "fancy" or combining units from memory— "counters," and "fixities and definites"—according to the law of association, and testing the results in some range of choice, rather than imagination that's infused with new perception.[63] (Henri Poincaré's century-old tale of mathematical creation is pure fancy.[64] "Atoms" are held in memory or a given cache—"motionless . . . hooked to [a] wall," they're taken down and put in "swing" to "flash in every direction," so that some of "their mutual impacts may produce new combinations." The "rules" for choosing atoms and their combinations, however, are "felt rather than formulated"— really, there are no rules just a feeling. Does this kind of intuition traverse fancy, for unknown regions where "to invent, one must think aside," or to put thinking aside, one must see askance?[65] Salvador Dali urges something similar in his "paranoiac-critical method" that risks the unplumbed depths of imagination, seeing in as many extravagant ways as possible, full of genius and folly alike.[66] Maybe shape grammars are Dali's dream and the paranoiac's delight. For Paul Valéry, poetic invention is also combination and choice, although unequally—"genius is much less the work of the first . . . than . . . the second." Today, there are updates for fancy in the syntax for symbols with its preestablished categories—the so-called "language of thought" that goes when the law of association fails. This marked increase in calculating, however, is no gamechanger—in mind or computers. The snag with fancy isn't how to combine things recursively—that's straightforward. The real problem is the vocabulary of counters—the units, atoms and the like in words, etc. needed for association and syntax to build on. In shape grammars, there are no units and so, no categories. Both are defined on the fly, time and again as they're put to use.) Some argue that fancy and imagination are arbitrary, a distinction without a difference—they vary only in degree, and not in kind. The distinction can be drawn either way in shape grammars, with marvelous indifference. (Likewise, it's no big deal to separate Poincaré and Valéry from Dali in shape grammars, or to keep a trio that performs fluidly as one.) Many in education, though, settle too willingly for visual analogies, in a misguided effort to make new perception unnecessary. Fancy and imagination are distinct, so that you see exactly what you're taught by rote, and never in your own way. It seems worth it, to get a good job. The hard questions are simply ignored—what are units for, where do they come from, and how do they change? Visual calculating in shape grammars answers these questions, and moves on. And finally, there's all of the technical stuff on computer implementations for shape grammars—in the cases where the elements in shapes are given in different analytic expressions, for example, linear or quadratic polynomials. This takes some math, but it's totally fascinating. It's amazing how much needs to go into computers to get them to see in the way that shape grammars do without visual analogies, for schemas and rules to work easily with parts that can change without rhyme or reason, with transformations of varied kinds, and with shifting boundaries. This is indispensable—until I actually try a rule, there's no way of telling what I'll see next. With embedding, I can put whatever I wish in shapes, and there's no trace (memory) of this if I decide to look again. Somehow, I've circled right back to Wilde's beautiful form and to von Neumann's Rorschach test—to insight and imagination lodged in the embed-fuse cycle

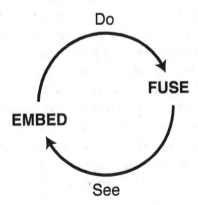

that keeps shapes and shape grammars forever vital. And can anyone doubt this—that it's the very heart of the matter? The conclusion is inevitable, yet to many it may shock and surprise—shape grammars are a more generous standard for calculating than Turing machines and computers. The proof is evident in art and design, and pictures and poems, and for Coleridge farther out than fancy goes in imagination's magical realm of esemplastic power.

**Q5.**

How is it continuing to develop? What is the future of shape grammars?

**A5.**

This asks a lot—there's no telling what I'll see in the embed-fuse cycle, or where this goes. For art and design practice and in education, schemas are key—defining them and applying them to exploit insight and imagination. Even alternatives to schemas and rules are possible to expand the sweep of established practice, still in concert with the embed-fuse cycle for visual calculating. And computer implementations will remain an important area of work, first for exotic elements (odd curves and crazy surfaces) and what they mean for embedding, and then for schemas and the assignments that fill in open variables to say when shapes look alike to try rules—are there measures, dynamic or not, for when shapes are copies, or examples of the same thing? But visual calculating in shape grammars is so agile that future work may strike anywhere, like lightning—why not something really far-out, something strange the way shape grammars were originally and may still be to many. It's not for me to know how shape grammars and their applications will evolve. It's for others to say, the heroic few who try them and test them out. And luck matters, as it does in shape grammars— I'm ready to be surprised. It will be great to see how this unfolds as time goes on. The trick is to be patient and to hang around long enough, or so my friends like to tell me. This can't be rushed—at least if Coleridge's imagination is any gauge (200 years and still waiting for "the order of our universes" to alter), or Thomas Kuhn's infrequent scientific revolutions. It may take a long time to see things anew, for the paradigm shift that accepts shape grammars totally, even in innovative, fast-moving times. Visual calculating changes everything in the flash of new perception. Of course, paradigm shifts happen all the time in shape grammars; they're part and parcel of how rules work when they're tried in the embed-fuse cycle—hence, neither numinous nor ever a mystery.

I don't need to cast about, helpless for new ways to go on; they're what I see now, in plain view—everything is superficial, never hidden out of sight, always there and ready to use. There's no secret underlying structure, no visual analogies that I have to learn, to interfere with the parts I find, recognizing merely a few special ones and barring the myriad that remain. Structure, it seems, is an especially invidious kind of censorship that's easy to affirm—everyone craves it, only to lose "the whole fiery-coloured world." This is pretty heady stuff, and a key aspect of seeing that's at the quick of art and design. It's the reason for embedding and why shapes fuse. This explains how rules exploit ambiguity as they apply to shapes to change them; it opens the way in fully, for insight and imagination. (To Wilde, the mystery of a beautiful form is superficial— "one can put into it whatever one wishes, and see in it whatever one chooses to see." Reason and thought are quick to dismiss the alluring surface and all that it shows for the deep and eternal truths in visual analogies—established descriptions and representations that take away the looking. But seeing turns away from logic and proof in true and false, and right and wrong, to embrace beauty and delight in unsettled appearance with vast and incalculable possibilities—"The true mystery of the world is the visible." Lord Henry makes no apology for this in *The Picture of Dorian Gray*—

> People say sometimes that Beauty is only superficial. That may be so. But at least it is not so superficial as Thought. To me, Beauty is the wonder of wonders. It is only shallow people who do not judge by appearances. The true mystery of the world is the visible [what the fickle eye sees/embeds each instant], not the invisible [the hidden/underlying structure the mind takes for granted, there once and for all].[67]

My copy of Wilde's story has the longer title, *The Uncensored Picture of Dorian Gray*. Victorian scruples against "writing stuff that were better unwritten" are why this addition is necessary, and the ambiguity in it highlights what I want to say. Beauty overtakes the visual analogies thought contrives. Maybe this structure is the natural way of the mind, but it blinds the eye—beauty is superficial, as long as there's more to see. Without underlying structure and units, only the superficial is left, in splendid profusion for art and design. In shape grammars and for Wilde, there are simply uncensored appearances—yours to choose. These change in the blink of an eye—for shapes and pictures, and for Dorian Gray, Lord Henry, et al. What was is only what is in memory and the mind, when things are objects with an underlying structure—"essentially fixed and dead.") It strikes me sometimes that going on and on about embedding and shapes that fuse is tiresome. But it repays the effort not to forget, especially when I'm not in the mood to see for myself, and I retreat too eagerly to visual analogies and rote structure. It's frighteningly easy to welcome dull fixity when others do, and they invite you to join them—there's comfort and security in numbers. Some of this intersects Coleridge—he talks about "that despotism of the eye," "the film of familiarity," and "the lethargy of custom." His words alone needn't have any particular meaning; their uses are many—now, for the kind of invariant seeing without seeing in STEM, that's taught in schools and tested rigorously, enforced for right and proper use. It's a function of imagination to overcome the film (cataracts) of satiated expertise and the lethargy of learning to see for the first time—time and time again. Is there any better reason to teach art and design equally in their own right, as art and design purely—not because some say they're instrumental and make it easier to teach other, more important stuff, but because they make it easy to see how to see, uncensored, with no definite purpose, without announced aims and goals? (This isn't always studio teaching, at least not in architecture.) Then, embedding extends farther than identity, past fancy, and the fixed units and counters in symbolic calculating, to modify seeing on the fly. I'd like to find an alternative way to go on, if only to talk about something else for a change—but no such luck. Embedding is implicated fully in whatever I try, whenever I look, in perfect repetition that ensures new perception throughout. Everything is vital in the pulse and the throb of the embed-fuse cycle; nothing stays the same once rules in schemas are put in play.

## Q6.

Do you know of any examples of where shape grammars have been practically used in architecture and design?

## A6.

I know of a respectable number of architects and designers who have used shape grammars in their creative/professional work, but they try to keep quiet about it, because the very idea of visual calculating is still too dangerous. The risk isn't worth taking, at least publicly, so I'm not going to say anything. It's poor marketing—the romantic rule for genius that it doesn't know its true method still holds sway in art and design. There's something curious about this, because shape grammars are a method that can always do more to exceed (transcend) anything that's expected or given. They should be used for what they're good at—to see and do in surprising ways, to go on like this with new perception. Isn't that how visionary genius works with insight and imagination? Visual calculating in shape grammars is meant to answer questions like this. No doubt, a lot remains to do—new questions seem ceaseless—but for me, this is where shape grammars and the embed-fuse cycle finally pay off and show their awesome power and potential. It's not so much what calculating does for art and design, rather it's what art and design do for calculating. Art and design inform calculating in shape grammars, so that shape grammars can reciprocate meaningfully in art and design—without loss, to see things as in themselves they really are not. This is something of a slogan or mantra for shape grammars. It's worth saying again and again, as a reminder that calculating in shape grammars is visual and not symbolic. My students insist they don't need reminding but anxiously apologize for ambiguity to betray their qualms—steering them past symbolic calculating (visual analogies) to seeing is never trouble-free. We all fiddle with symbols to calculate, and many of us find this seductive and satisfying—that's what calculating is and how it's supposed to be. I guess it's natural to describe what you're doing as a symbolic process, typically with words (atoms and units), without ever noticing that it really is not. It's easy to match what you've got with a mathematical model or structure that's simply taken off the shelf, and to stick with it—it can't be wrong, it's proven in solid mathematics. Everyone does it that way. You can talk about an underlying set of parts as a graph or tree, or mappings and isomorphism—some are positive that this is how painting and poetry work, contentless because they're stripped of ambiguity and imagination. It sounds good to say that you've found an isomorphism; the result is remarkable to behold—you're proud of what you've accomplished, and confident of what your method explains and that nothing is lost, and your friends are always impressed. Maybe this kind of structure is implied in the words we use to grasp what we see; maybe it mimics the hierarchical (tree) structure of language itself, to organize seeing in terms of what we know how to say. Maybe it's simply all we can think up, tied to recursion, words, and Chomsky's merge—bound to combine atoms with lawlike mappings.[68] The critic especially must guard against this alluring trap and its empty effects—it's a convenience to avoid. Is it any wonder that he/she succeeds only as an artist? But this isn't always so easy. Overarching structure is stressed in everything we're taught—everyone agrees that it's perfect, tried and true. There's no reason to ever tamper with any of it. I've already said as much for new perception— the reason for school is to limit your options as you memorize units and count, and to recite what's expected on tests and on the job. It's hard if not impossible to forsake everything you've learned and to break old habits, to abandon shared standards and the safety and comfort of symbols for hopelessly ambiguous shapes—to ignore rote divisions and specialized parts, to see on your own not tethered to visual analogies, to trust your eye no matter where it goes with the end never in sight. It takes true grit to follow through on this, every time in a fresh start. Insight and imagination may seem fleeting and hard to grasp—in fact, they last no longer than the twin beats in the embed-fuse cycle in shape grammars. But insight and imagination aren't infrequent—they're present all the time, impossible to escape in visual calculating. In every rule that's tried, the pulse is constant in bursts of esemplastic power—embed-fuse, embed-fuse, embed-fuse, etc. It's the risk you run to go on, always prey to something new. It's never safe—surprises are everywhere. In art and design, and for the critic as artist, too, the eye's search for security is senseless—the search doesn't matter because the eye doesn't see. (Heuristic search in AI fails for the same reason—there's nothing to see. Whatever design is, it isn't search in the security of a design/search space of given descriptions and representations, paradigmatically in visual analogies.) Going on is one risk that's a risk not taking.

## Q7.

What is the typical reaction of designers to the idea that their forms can be broken into grammars? How do they feel about this possibility?

## A7.

When I first started out nearly 50 years ago, the reaction was automatic—"Are you crazy? You must be nuts." But as shape grammars were shown to do more and more, the reaction turned into something defensive like this—"Well, no matter what you can see and do with shape grammars, there's a lot more to my designs than that. And there's always going to be more, much-much more. Shape grammars are limited—you misunderstand art and design, whenever you equate creative genius and method. Genius is unbounded; it exceeds all method. There's no way to describe my designs completely. I can put into them whatever I wish, and see in them whatever I choose to see." Yes, we agree completely, that's exactly how it is—that's the perfect response for shape grammars, because they ensure that art and design are totally open-ended. Every one of us senses the world variably to make it meaningful in a personal way—seeing, hearing, touching, etc. now and then again, in an ongoing process. Shape grammars encourage this and show how it's practicable in calculating. That's what shape grammars are for, with schemas and rules in the embed-fuse cycle. And certainly, I would like to see more artists and designers use shape grammars in their creative work, and to teach art and design always with insight in mind and imagination first and foremost. Both insight and imagination are key whenever seeing and sense experience are valued above all; they should be nurtured and practiced explicitly as the focus of art and design, and infused throughout STEM to keep it vital. Shape grammars open up art and design to show how insight and imagination work. It's a marvelous way to see what they do, and to try them out and use them in this way and that, at will in any way you please—in effect, for everyone to experience genius firsthand. Grammatical anaphora can be a dangerous thing—what does "they" refer to in the previous sentence? But my ambiguous antecedents are for the most part intentional. Sometimes, it seems better to explain more reciprocally, by not trying to be completely clear—when things overlap, there's no reason to write about them distinctly. There's more in them than I can say. This time, it's to highlight a new kind of rule that summarizes what I've been trying to show—

shape grammars → insight and imagination

And the inverse of this rule may work equally, as well—just flip its left-hand side and right-hand one. The esemplastic power replete in insight and imagination is key for visual calculating in shape grammars—it's the motive force that puts the embed-fuse cycle in motion, and it's multiplied in its every pulse. And now when things involve calculating, it makes pretty good sense to take them seriously—or at least, to pay close attention.[69]

**Coda.** Everyone is worried today that computers and AI will finally learn to do everything that's useful, and thereby make work and employment hard to find in the foreseeable future, and our sought-after professions irrelevant and unnecessary. And without new perception, this is probably true. But then, what's left for you and me to do? I guess it's our destiny to be redundant. Once this happens—maybe it already has to a few of us—we'll have to find useless things to fill our free time. One place to look is art and design. (My list in A4 extends a little farther than this, to include "the humanities, the arts, physical fitness, and other subjects that don't divide everything into units." But maybe I should add a little more—it's easy to put in literature and criticism explicitly, then fashion design is a must, other capricious and ephemeral pursuits, and what we care about beyond instrumentality.) Wilde takes this very seriously in his usual, extravagant way. "All art is quite useless"—

> The only beautiful things, as somebody once said, are the things that do not concern us. As long as a thing is useful or necessary to us, or affects us in any [definite] way, either for pain or for pleasure, or appeals strongly to our sympathies, or is a vital part of the environment in which we live, it is outside the proper sphere of art.[70]

Computers and AI use prior data to learn useful things or about them. (Nowadays, this would be the preferred way to define visual analogies, solely in terms of given training sets, without regard to von Neumann's axiomatic/verbal approach to description in which, say, a triangle has three edges, etc. It spells the end of description in McCarthy's kind of AI, and a full return merely to discrimination.) Art and design are indifferent to this; there's no prior data that computers and AI can use, and so, nothing to learn. What comes first in art is art—pictures from pictures, and also from tree-trunks, clods of earth, and clouds, whenever perception is new. The trick is in Wilde's critical spirit, in the aesthetic formula to see things as in themselves they really are not—things are useless if use doesn't matter and isn't a concern. (Architecture shows how this works. Delight is key as it interacts with firmness and commodity—the two are necessary yet for the time being, they go largely if not totally unnoticed, or noticed in a strange and wonderful way.) It seems that things are useful as in themselves they really are, and useless as in themselves they really are not—only useless things are beautiful forms. Visual calculating in shape grammars fosters uselessness and makes it something to embrace. In shape grammars, data is retrospective and not required to go on—in fact, it leaves no permanent trace or lasting memory. I can do what I see even if it's not what I've drawn. With embedding, things are always ambiguous and open-ended; their meaning depends entirely on how I divide them now—in the present, at this moment, independent of their past history or what I think I've done. Every shape is a beautiful form or the Rorschach test. (And aren't questions like the Rorschach test, too, including the seven I've just answered?) Shapes aren't units or visual analogies I can store in computers or keep in my head, to recite by rote whenever they're needed. Shape grammars change useless things with useless rules. (All rules are useless, as identities in the schema $x \rightarrow x$ that keep shapes the same in infinite loops, and then more importantly, because descriptions of rules vary all over the place, as the schema $x \rightarrow \Sigma\, F(prt(x))$ shows when different terms $F(prt(x))$ fuse in a single outcome—use is elusive, embedding makes it hard, if not impossible, to tell what rules do before additional rules are tried, what happened in the past depends on what happens now.) The embed-fuse cycle pulses time and again with new perception, to make insight and imagination, and art and design possible. Wilde is keen on Socialism to release this creative energy and vitality, and to sustain it fully.[71] Today, he would embrace AI and machine learning eagerly, as the logical extension of this—Socialism with all of its administrative schemes and plans must end with computers for maximum efficiency, matchless expertise, and perfect utility, and above all, for fairness that's free of bias, and the ambiguity and noise that engender nagging doubt.[72] AI will do useful things, to provide free time for visual calculating in shape grammars, for art and design, for useless things not tied to prior data and learning, for what isn't AI—isn't that why the Rorschach test absorbs von Neumann? Is this something worth fostering, or is it simply a trivial fraction in an inescapable technology, an infinitesimal that will fade to zero? Many are resigned to the latter if not committed wholeheartedly—that sometime soon, computers and AI will reduce everyday experience to prior data and learning, to make it ever so easy to settle for convenience and utility, for visual analogies and other descriptions to take the place of original experience that's intuitive above all—maybe this will be the golden age of education. An old friend of mine is sure that AI is the open sesame to a glorious utopian future, even as he questions Socialism and the reasons for its administrative choices and decisions. This is most likely a passing concern, as AI replaces Socialism more and more to restore confidence and trust, and to ensure that all doubts and fears disappear in coherence and opacity. AI grasps the eternal and unchangeable as Plato's philosophers do, and keeps oddly silent when pressed for details of how this works and why—there are no reasons to probe.[73] AI is a black box, bulging with details and qualifications in figures, and data and statistics—"that cowardly concession to the tedious repetitions of domestic and public life."[74] It's about numbers and measurement, and the facts and nothing but the facts—in a quest for intelligence that's entirely averse to insight and imagination, and strictly banausic as a result. Reasons are for free time only, useless things for gossip and philosophy alike—it makes no difference. (The current purpose of AI is to do everything it tries well, without trying to understand it, and so to become what needs to be understood.[75] Maybe computers will explain computers, with data and statistics. Is this the true nature of scientific progress?[76]) Of course, Plato presents a stark dichotomy. The unchangeable and the uncertain needn't be related, in the way Socialism and AI are supposed to support the creative and vital in the full-flux of new perception. This is an unlikely relationship at best—maybe it will work, and maybe it won't. Sooner or later, any relationship may put the unchangeable and the uncertain at odds—even irreconcilably. However this turns out, there's Plato's uncharted region of the many and variable to traverse freely, wantonly self-reliant with neither map nor guide, armed with aesthetic imagination only—wandering aimlessly in pursuit of who knows what in an endless and ever-changing maze, at constant risk of something strange and new to charm the eye. In these inconstant precincts with their borderless environs, the untethered (unstructured and memoryless) eye is quick to find more to see. Insight and imagination animate the critical spirit and put everything in motion, exciting pleasure and delight in "the whole fiery-coloured world." Without insight and imagination, and the critical spirit to light (find) the way, this isn't a particularly agreeable place for computers, even vast and super-fast; maybe AI and computers will provide the free time and leisure to get there, but they won't help to stay—

> The universe increasingly has a common technology, and in time may constitute one vast computer, but that will not quite be a culture.[77]

My approach to computers and symbolic calculating, throughout my seven questions and answers, has been mostly as critic/artist—on the one hand, dismissive, and on the other hand, inclusive.[78] Insight and imagination subsume and exceed fancy in visual analogies (descriptions)—in visual calculating in order to bring in art and design. Computer descriptions in words, in numbers, in data and statistics, etc. limit what they describe, so that experience is final, and things (objects) are fixed and dead. Computers and AI "will not quite be a culture," as long as there's a beautiful form and the critical spirit is free to go on, to fuse and re-divide—then no description is ever complete or immune to re-vision. Ambiguity and new perception are rife with contradiction, discontinuity, uncertainty, and change; they put things in motion and keep them vital; they swamp AI and its store of constant data with half-knowledge. This much is sure—in shape grammars, there's always more to see that hasn't been seen before. Schemas and the rules they define embrace surprises in fickle plans and clever tricks for shapes and retrospective relationships in graphs, hierarchies, mappings, topologies, and descriptions and representations of many kinds. Seeing is uncensored, never a freedom too far; structure is fleeting and forever in flux. Shape grammars are strange and wonderful—without prior data and learning to frame and map out what the future holds, rules in the embed-fuse cycle reconfigure/redescribe the past in terms of a present they re-create freely as they're tried. Shape grammars aren't about given descriptions and representations in axioms or data and learning; they're about ambiguity in beautiful form, as calculating goes on and on. Shape grammars are as in themselves they really are not; they're useful (indispensable) because they're useless. Shape grammars are beautiful things.

## EXHIBIT 1: THEORY

Shape grammars calculate with rules in a recursive process; they have a key property Turing machines don't—and this goes for like methods of calculating, such as generative (formal/symbolic) grammars. Shape grammars are open-ended (unbounded) in two ways—their rules (1) divide shapes into an infinite variety of parts, as they (2) produce an infinite variety of shapes. Turing machines and generative grammars allow for (2) and not (1)—in shape grammars, parts aren't determined (described/represented) before calculating begins but vary freely as rules are tried. This makes Turing machines and generative grammars a special case of shape grammars.

In a shape grammar, a rule

A → B

is made up of any shape A in its left-hand side, and any shape B in its right-hand side. Maybe one bunch of lines is changed into another bunch

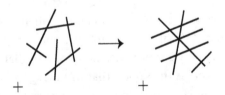

But things are usually less haphazard—for example, when a square is changed into a triangle

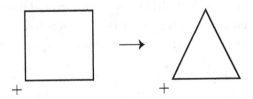

Either one of the shapes A and B can be the empty shape or a blank space

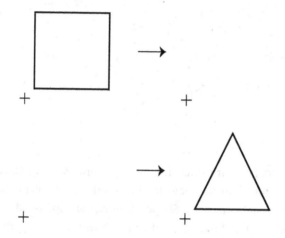

or both A and B can be the empty shape

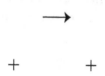

It seems that just the idea of change (action/movement) in an arrow (→) is fine when it comes to a rule.

The rule A → B applies recursively to any shape C whenever there's a transformation t that makes A part of C; that is to say, t(A) is a copy of A that's embedded in C. The transformation t may be a Euclidean isometry or a similarity, linear, or wild and crazy. Whatever t is, the goal is the same—that t(A) looks like A and is something in C. This usually goes for similarity transformations, where orientation, place, and size don't matter, but even this may be dicey for b, d, and p without added serifs, and for + and ×. It's up for grabs for other kinds of transformations; for one thing to look like another is a fickle match that the transformations do their best to capture. I'm apt to stick to the similarity transformations to avoid strife most of the time, but in general, there's no reason to be conservative—what the eye feels has the final say, whatever this means for the transformations. The relationship between t(A) and C is given in the expression

Exhibit 1                                                                 61

$$t(A) \leq C$$

where $\leq$ denotes the part relation defined in terms of embedding. The result of applying the rule $A \rightarrow B$ to the shape $C$ is another shape $C'$ given in the handy formula

$$C' = (C - t(A)) + t(B)$$

Simply put, the copy of A is taken away from the shape C to produce the shape $C - t(A)$; then the corresponding copy of B is added to $C - t(A)$ to get the shape $C'$ that succeeds C—the idea is that B is copied in the same manner as A, although again, there's plenty of room to fool around here in the way $t(B)$ is produced and combined with $C - t(A)$. Everything fuses in $C'$—both $C - t(A)$ and $t(B)$ are assimilated completely, leaving no trace of either. (I can't say it often enough. Singly, the shapes $C - t(A)$ and $t(B)$ are unanalyzed without distinguished parts—this goes for every shape, including the sum $(C - t(A)) + t(B)$ in which the parts $C - t(A)$ and $t(B)$ disappear.) Whenever the rule $A \rightarrow B$ applies, it recapitulates the embed-fuse cycle in two stages—(1) finding a copy of A in C, and (2) replacing it with a copy of B. Stage (1) provides an analysis of C, dividing it into two parts $t(A)$ and $C - t(A)$, and stage (2) provides the corresponding synthesis for $C - t(A)$ and $t(B)$. In shape grammars, analysis and synthesis are part and parcel of a single recursive process—each precedes, or equivalently, follows the other. But this is evident in the augmented diagram for the embed-fuse cycle

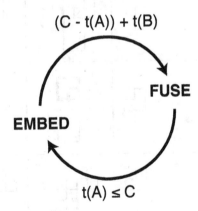

in which embedding is confirmed in $t(A) \leq C$, and distinct pieces fuse when $t(B)$ is added to $C - t(A)$. How analysis and synthesis unfold in parts ($\leq$) and sums (+) as I calculate with shapes is illustrated in table 1 in a kind of self-reproducing mitosis, that splits an initial cross

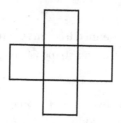

that's unanalyzed into two or more replicas of itself, that decrease in size by one half in an ongoing geometric series. This process plays out freely in terms of squares and L's

**Table 1**  Calculating recursively with shapes and rules in the embed-fuse cycle.

| C | A → B | t(A) ≤ C | C − t(A) | (C − t(A)) + t(B) |
|---|---|---|---|---|
| | | | | |
| | | | | |
| | | | | |
| | | | | |
| | | | | |

**Exhibit 1**                                                                                63

that appear and disappear as rules are tried; there are two rules

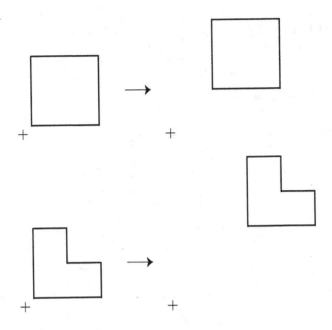

that are both simple translations on the diagonal axes of a square and an L-shape. The transformations t to apply the rules are similarities—translation, rotation, reflection, and scale, and their compositions. Copies look alike in an uncontroversial way, in shapes and rules. (Flipping the sides of rules in inverses provides origins for the initial cross going back in time. This is explicit for the second rule, the first is its own inverse. And if my replicas look too small to divide, put in a rule to make them bigger—to grow offspring into productive adults, at once or in measured spurts. Easy, and easier for schemas in which scale/size varies continuously.) As the rules are tried, squares and L's turn into one another, highlighting the embed-fuse cycle. Changes in analysis that switch between different parts affect what happens next in synthesis—I do what I see. Analysis is double-edged—it's key for synthesis in the embed-fuse cycle, and its impact shows cumulatively in the ongoing synthesis (generation) of shapes. Even rules that only move parts around like game pieces (squares and L's) do pretty amazing things, when parts aren't fixed and constant but fuse and re-divide. There are untold surprises with the same rules and fickle pieces. Table 1 does the trick, if you can split the initial cross into smaller crosses, so that their numbers grow in sundry places and spatial relations. Or see/try something else with additional rules for squares and L's, or other parts.

Rules can be generalized usefully in terms of schemas. A schema

x → y

is a pair of variables x and y separated by an arrow; x and y take shapes as values in an assignment g, so that

g(x) → g(y)

is a rule. If g(x) is the square

and g(y) is the triangle

then the rule is

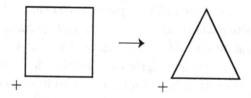

In summary, the schema x → y applies to a shape C when there's an assignment g and a transformation t, such that

t(g(x)) ≤ C

This defines the new shape

(C − t(g(x))) + t(g(y))

**Exhibit 1**                                                                          65

Sometimes there are nice relationships between the values assigned to x and y. For example, the two rules used in table 1 are in the schema x → t(x) that defines transformations of x—in particular, diagonal translations. Such relationships are discussed in Exhibit 2 and Exhibit 3.

Of course, what's missing in all of this is a full explanation of shapes, the part relation ≤, and the operations of sum + and difference −. In general, shapes and the rest depend entirely on what I see—things in tree-trunks and clods of earth, and pictures, as well, that I can change, taking away and filling in. But to show how the math works, some initial restrictions are necessary. To start, shapes are finite throughout—they can be made in a finite region, maybe drawn on a piece of paper in a finite number of pencil strokes or other actions, and include a definite number of elements. And to go a little farther, a shape is any finite number of basic elements—points, lines, planes, and solids of dimension 0, 1, 2, and 3, respectively. The shape

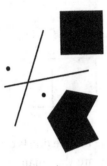

is made up of points, lines, and planes. (I've left out solids only because I can't fit them on this page—a drawing of a solid, say, a cube

isn't a solid. In my drawing, it's five lines and a plane on a flat surface.) Basic elements have the properties given in table 2. In particular, the boundaries of basic elements—points for lines, lines for planes, and planes for solids—are finite; in fact, the boundaries of basic elements are also shapes. To continue in this vein, the basic elements in shapes are assumed to be maximal, that is to say, no two of the same kind—a line and

**Table 2** Basic elements and their properties.

| Basic element | Dimension | Boundary | Content | Embedding |
|---|---|---|---|---|
| point | 0 | none | none | identity |
| line | 1 | two points | length > 0 | partial order |
| plane | 2 | at least three lines and finitely many | area > 0 | partial order |
| solid | 3 | at least four planes and finitely many | volume > 0 | partial order |

a line, a plane and a plane, etc.—can be combined to make a third, that has the two original elements embedded in it. The pair of squares

is five lines

not seven, or maybe eight when the middle line is counted twice, once for each square. The maximal elements in a shape are the smallest number of biggest elements that make the shape—in my example for two adjacent squares, the five longest lines, a top horizontal and a bottom one, and three verticals left to right. These are the lines that someone skilled in drafting would draw with a T-square and triangle. Distinct points are always maximal

lines are maximal whenever they're collinear and separated by a gap

**Exhibit 1**                                                                                          **67**

or not collinear, as in these three shapes

and similarly for planes

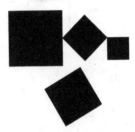

and solids. Algebras of shapes fill in the other details—algebras are defined for the part relation ≤ and two operations for sum + and difference −, and all of this is augmented with transformations. The dimension i of each kind of basic element determines a separate algebra $U_i$, for shapes of dimension i. These algebras are ready to combine, so that the shapes in each are independent in the same space, although related to one another geometrically. In this shape

two points are the endpoints of a single line for three basic elements, but the points aren't parts of the line—if I take the points away, the same line is still there, and vice versa. Points, lines, planes, and solids are separate kinds of things; they're never parts of one another. Instead, boundaries provide mappings between shapes of different dimensions—the properties of basic elements ensure that the boundaries of shapes are shapes. The boundary of a shape of dimension i + 1 traces its limits in a shape of dimension i, without being part of it. For example, the shape made up of a pair of square planes has a boundary of six lines

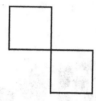

and these are bounded by six points

Algebras of shapes can be indexed in more detail in terms of the dimension j of the space in which transformations are defined for shapes; this gives the series of algebras

$$U_{00} \ U_{01} \ U_{02} \ U_{03}$$
$$U_{11} \ U_{12} \ U_{13}$$
$$U_{22} \ U_{23}$$
$$U_{33}$$

For basic elements of dimension i, i ≤ j. In the case where i = j, a point is defined in a point, but of real interest, lines are defined in a line

**Exhibit 1** 69

and planes in a plane

along with solids in a solid. (Sometimes, sculptors are good at seeing solids in solids—Michelangelo's *Prisoners* may show what happens when there's a mistake.) For basic elements of multiple dimensions, j is always greater than or equal to the biggest dimension. In this shape

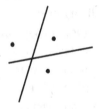

with points (i = 0) and lines (i = 1), j is 2. The part relation ≤ and the operations of sum + and difference − define a generalized Boolean algebra—with the exception of $U_{00}$ that's a standard Boolean algebra for a single point and the empty shape. (This is the same as the Boolean algebra for 0 and 1, or true and false.) When i = 0, the algebras are atomic—shapes that have points have a finite number of them, and a finite number of parts. Moreover, the empty shape is the zero in these algebras, but for j > 0, there's no universal shape, as this involves indefinitely many points—and there are no complements for shapes, as well. In contrast when i > 0, no algebra is atomic—shapes that have elements (maximal lines, planes, or solids) have a finite number of them, but indefinitely many parts. Again, the empty shape is the zero, there's no universal shape, and shapes are without complements. (The shapes in my algebras $U_i$ can be represented as "Stone spaces"—but for i > 0, these are infinite sets outside of calculating/drawing. My algebras also overlap with "individuals" and "mereology"—with a zero and without infinite sum. Both of these differences are key for shapes, although especially the finite restriction on sum.) The embedding relation explains all of this; it's a partial order that can be used to define parts, and sum and difference. If shapes are made up of points, it's a chinch in the old New Math. Parts correspond to subsets, and sums correspond to set union—distinct points in different places don't fuse and identical points in the same place are included only once. And difference is relative complement. (Spatial relations are exactly the same—when they're combined and used to calculate, the shapes they contain are like points. As a result, shapes in spatial relations are 0-dimensional, even if they're not. Many designers balk when I tell them that "3-dimensional" designs are really 0-dimensional, if they're only made up of elements that are fixed and independent like Lego bricks—entirely fancy in spatial relations, particularly in computers for parametric design and BIM. Being 0-dimensional is everywhere in symbolic calculating; it goes all the way to Turing machines—and this is a ready way to prove that Turing machines, generative grammars, etc. are a special case of shape grammars. I like to think that design and calculating will be different somehow as soon as the embed-fuse cycle kicks in for seeing, and that "the order of our universes will have been changed"—although I'm easily convinced that such aspirations are hopelessly romantic, if not fantastic. Remarkably, ordinary lines, 1-dimensional basic elements, show the extraordinary change from identity for points to embedding in visual calculating. And lines are super in other ways, as well; they're a marvelous technology in the algebra $U_{12}$ with only pencil and paper and commonplace drafting tools, and key in Leon Battista Alberti's definition of architecture, apt today nearly six centuries later. Nonetheless, the designers I talk to—and many artists—are keen on computers and mistrust shape grammars and their insistent vagaries. Computers strike a Faustian bargain; they promise to facilitate design but take away the looking, to repeat endless variations on demand only to forgo imagination and so forfeit the content of the soul. This isn't an offer you can't refuse. Be bold—the limits of computers, and the parametric design and BIM they foster, are obvious to anyone itching to see.) So far, so good—but for lines, planes, and solids, reduction rules are a better way to frame definitions for parts, sums, and differences than embedding alone. Let's start with the reduction rules for sum; they're given for lines in table 3. There are three separate cases—(1) for embedding, (2) for overlapping lines, and (3) for discrete lines. (The fastidious and rigorously inclined are apt to ask for more. In particular, overlapping lines and discrete lines can be defined entirely in terms of embedding. In fact, the three relations are mutually equivalent—going from overlapping lines or discrete ones to embedding is a nifty trick. Embedding makes intuitive sense for shapes and shape grammars, to follow the eye in art and design; in logic and philosophy, Nelson Goodman prefers overlapping, but together with Henry Leonard, he opts for a calculus of individuals that are discrete and not. Such similarities

Exhibit 1                                                                71

**Table 3** Lines and their reduction rules for sum.

Assume lines are ordered by their endpoints, and let l and l' be any two of these lines. Then, lines are changed according to three rules. These are used recursively in any order until no rule applies.

(1) If l is embedded in l', either like this

so that they share an endpoint, or possibly two, or l and l' are embedded like this

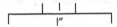

so that there are no endpoints in common, then remove l.

(2) If l and l' overlap but neither is embedded in the other

then replace both lines with the line l'' fixed by the leftmost endpoint of l and the rightmost endpoint of l'. This is the longest line with an endpoint of l and an endpoint of l'.

(3) If l and l' are collinear and discrete, and share an endpoint

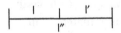

then replace both lines with the line l'' fixed by the remaining endpoints of l and l'.

are nice, even with the differences I've already mentioned. Still, these are crucial and bear repeating—for Goodman and Leonard, individuals have infinite sums and are never empty; for shapes it's the opposite for both.) Many find case (3) a surprise, that two discrete lines can make a third—after all, they have no content in common, only a boundary (limit) that's a point and not a line. To see how the three reduction rules work in concert, let's suppose that I add four squares

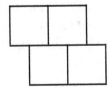

in this zigzag sequence

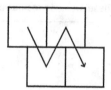

Then, the reduction rules are 2; 1, 2, 3; and 1, 2, 3. And if I try the alternative sequence

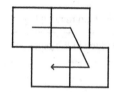

that goes another way in a boustrophedon scan, the reduction rules are 1, 3, 3; 2; and 1, 1, 3. Wherever the eye goes, there are the same nine maximal lines—I guess eye tracking is mostly irrelevant. And this extends immediately to embedding and parts. If A + B = B, then A ≤ B—intuitively for lines drawn on a piece of paper, if I trace the maximal lines in A and put them on top of the maximal lines in B, A matches (aligns with) parts of B and disappears. In keeping with embedding, A = B whenever A ≤ B and B ≤ A. For planes and solids, things are pretty much the same—the tripartite format for lines in table 3 works for planes and solids, as well. Nonetheless, the reduction rules for sum are somewhat harder to define than the ones for lines—the boundaries of planes and solids are more complicated than the pairs of points in the boundaries of lines. In general, the reduction rules for sum for lines, planes, and solids turn any finite set of basic elements—each basic element is a shape by itself—into a matching set of maximal elements. And with all reduction rules for shapes, there's a kind of confluence—in whatever way the reduction rules apply, they end up with one and the same result. My example with the lines in four squares that sum in distinct sequences provides some convincing evidence—and the line by line proof isn't that hard to work out. The embedding stage of the embed-fuse cycle, $t(A) \leq C$, is effectively encapsulated in reduction rules for sum. In general, I denote them by $R^+$, maybe with some minor bookkeeping in indices i that specify basic elements by their dimensions. To test for embedding, simply check the equality

Exhibit 1                                                                 73

$$R^+(t(A), C) = C$$

for $t(A) \leq C$. To finish up the mechanics of my algebras, there are reduction rules for difference $R^-$. These are easier than $R^+$—with cases only for embedding and overlap. Discrete elements aren't involved in difference—for one minus another, even with overlapping boundaries, there's nothing in the second to take away from the first. The reduction rules for difference are enumerated in table 4; once again, this is for lines.

Neatly, parts are defined in $R^-$ just as they are in $R^+$. For two shapes A and B, if A − B is the empty shape, then $A \leq B$—no part of A is anywhere outside of B. In terms of $R^+$ and $R^-$, the fusing stage of the embed-fuse cycle, $(C - t(A)) + t(B)$, in which shapes fuse as they combine in sums and differences to assimilate $(C - t(A))$ and $t(B)$, is entailed in the composition that goes like so—

**Table 4**  Lines and their reduction rules for difference.

Start with any two shapes with the maximal lines in each ordered by their endpoints. Let l be a maximal line in the first shape and l′ be a maximal line in the second. Then the first shape is changed recursively according to three rules that apply in any order until none does.

(1) If l is embedded in l′

then remove l from the set.

(2) If l′ is properly embedded in l, so that no endpoint is shared

then replace l with the lines l″ and l‴ fixed by the two leftmost and the two rightmost endpoints of l and l′. Alternatively, if l′ is properly embedded in l, and there is a common endpoint

then replace l with the line l″ fixed by the remaining endpoints of l and l′.

(3) If l and l′ overlap but neither is embedded in the other

then replace l with the line l″ fixed either by the two leftmost or by the two rightmost endpoints of l and l′, so that l″ isn't embedded in l′.

$$R^+(R^-(C, t(A)), t(B))$$

And of course, the upshot of this is that the maximal elements in $(C - t(A))$ and $t(B)$ needn't be maximal elements in $(C - t(A)) + t(B)$. Maximal elements are preserved only if they're points, when they're like the members of sets. Four maximal lines make up a square, but just nine are needed to fuse the four staggered squares in the previous example—three horizontals and six verticals. (In the top zigzag sequence, adding maximal lines by their numbers gives 4, 4 + 4 = 7, 7 + 4 = 8, and 8 + 4 = 9; in the boustrophedon scan, 4, 4 + 4 = 5, 5 + 4 = 8, and 8 + 4 = 9. Try it for both sums and differences for the shapes in table 1.) The algebras I've defined for shapes are merely a start—they can also be extended with labels (symbols) and more inclusively, to add weights for colors, materials, etc. that incorporate a gamut of physical and sensory properties. And I can go on to other things in math, as well—for now merely to give the flavor of this in three examples, with scant attention to detail. First, registration marks show how transformations work for shapes and rules in algebras, to distinguish determinate and indeterminate rules, where the former apply in finitely many ways, and the latter don't without ad hoc restrictions or personal intervention. This is good for an equivalence relation between shapes—two are equal with respect to transformations (their unique geometry) if they have the same registration marks. In fact, it's a nice start to compare and order shapes in telling ways in art and design. Second, La Grange's theorem counts the distinct results defined in a rule $A \rightarrow B$, in terms of B and how it divides the symmetry group of A into cosets—what a rule does depends on how the shapes in its left-hand and right-hand sides are related. And third, continuity in rules involves topologies (structure), closure, and homomorphisms. If a rule $A \rightarrow B$ applies under a transformation t to change a shape C into a shape C', then $t(A)$ is closed in the topology of C; that is to say, the analysis of C with respect to the rule makes sense. Moreover, there's a mapping from every part x of C to a part of C', maybe $h(x) = x - t(A)$, such that the image of the closure of x in the topology of C is part of the closure of the image of x in the topology of C'—no division that's OK in C' implies a division not already in C. This extends the reciprocity of analysis and synthesis in the embed-fuse cycle in a retrospective analysis of C, so that its topology is consistent with the topology of its successor C'. Analysis and synthesis don't prize apart when rules are used to calculate—the former follows the latter, as shapes are produced recursively. Shapes and shape grammars connect with math in a host of marvelous ways in algebra, analysis, geometry, topology, etc.—sometimes, however, this seems rather scattershot if not entirely arbitrary. In truth, shapes and shape grammars are an area of math in their own right, tracing a separate arc through the different relationships for embedding and transformations

Exhibit 1                                                                                            75

that put seeing in calculating—the two go together seamlessly for rules in the embed-fuse cycle. Embedding, in particular, does the trick for visual calculating with shapes in shape grammars—for the special math that plays out abundantly in insight and imagi-nation, in the fickle perceptions of artists and designers lodged in drawings in fine art and design, in paintings, sculpture, etc. There's solid proof in seeing.

All of the math and plenty more is elaborated in part II of *Shape*, "Seeing How It Works"—G. Stiny, *Shape: Talking about Seeing and Doing* (Cambridge, MA: MIT Press, 2006), 159–310. *Shape* extends my original approach to shapes and shape grammars in G. Stiny, *Pictorial and Formal Aspects of Shapes and Shape Grammars* (Basel: Birkhäuser, 1975). This includes embedding, reduction rules for sum and difference, transforma-tions, registration marks ("points of intersection"), composite shapes in Cartesian products, operations on shape grammars and the languages (sets) of designs they define (surprisingly, this is missing in *Shape*), and how shape grammars and Turing machines are related to one another. *Pictorial and Formal Aspects* is the first place that I know of in which the embed-fuse cycle unfolds formally—if only implicitly—for use in visual (pictorial) calculating. It's taken more than two centuries to see that Samuel Taylor Coleridge's imagination and esemplastic power in the *Biographia Literaria* have a locus in math—for calculating in shape grammars that leave nothing to the imagina-tion when it comes to how rules are defined and apply in algebras of shapes, and for calculating that includes Turing machines as a special case. Really though, who would dare to think of such an outlandish thing—to run calculating through imagination in order to add to the soul? John von Neumann sticks to symbolic (computer) calculating in the standard way with descriptions and representations—fancy in visual analogies, spatial relations, parametric design and BIM, and like kinds of structure in statistics and prior data—only to question the completeness of this for pictures and the Rorschach test. It seems that seeing overflows with ambiguity, fickle and relentlessly open-ended in the ongoing aesthetic (perceptual) experience of the artist and critic alike—and gen-erally, in the personal experience of all of us. But shape grammars provide a creative alternative in visual calculating, one that "shows us the whole fiery-coloured world" in pictures and the Rorschach test, and word for word and letter for letter in Oscar Wilde's beautiful form—"The one characteristic of a beautiful form [Rorschach test] is that one can put into it whatever one wishes, and see in it whatever one chooses to see." This supersedes von Neumann's unresolved concerns in Wilde's fantastic aesthetic/critical formula, to see things as in themselves they really are not—seeing as the eye swerves in unimpeded flight, instead of counting 1, 2, 3, . . . with 0's and 1's. There's only one way to see things as in themselves they really are, and indefinitely many ways as something else. That's why computers run amok; a single way is never enough—even an eternal truth, or verified in "ground truth" with empirical certainty. No matter how compre-hensive (long) descriptions and representations are, whether in axioms or prior data, almost everything is left out in endless surprise—a triangle is three edges and then sud-denly, it's something else, like in a picture or the Rorschach test in individual percep-tion. The trick is to see more without anticipating it, and then to do what you see—as the trio of examples in Exhibit 2 shows, going from pinwheels that rotate in strange and spooky (impossible) ways to squares that really aren't squares, and back again to Wilde in a tragic kind of Pythagorean free play with squares and right triangles, com-mensurable only in their ambiguity. This is calculating beyond fancy in imagination's magical realm. The math proves it, as the eye swerves in schemas and rules in the embed-fuse cycle—relaxed on the one hand in scant summary in this exhibit, and more rigorous on the other hand in the fine and luxuriant detail of *Shape*.

## EXHIBIT 2: OBSERVATIONS

The best way to understand how visual calculating works in shape grammars, and why this isn't symbolic calculating, is to see what rules do when they're put to use. The left-hand side of a rule

A → B

is key; it tells what to see in any shape C to which the rule applies. The shape A is an example that can be copied in some way to embed in C. Typically, this transformation is a similarity but it may be linear or something more general. The right-hand side of the rule tells what to do; B is usually an example of a shape that can be copied in the same way as A, to replace the copy of A that's embedded in C. Or B might be an instruction to make marks in a specified way or to throw paint as some artists urge, at least from the Renaissance on. Whatever it does, once the rule is tried, everything fuses to complete the embed-fuse cycle

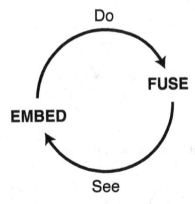

so that another rule, possibly the same one, can be applied from scratch, with no history/memory of anything that's happened before. I do what I see—right now.

This exhibit has neither a premise nor a conclusion; it's wayward observation without a definite argument—like Samuel Taylor Coleridge's thesis, antithesis, and synthesis/indifference, calculating in sync with the embed-fuse cycle that starts with an undivided (unanalyzed) shape and ends in the same way with an undivided shape. The exhibit consists mostly of different ways of looking, evanescent in examples that show the necessity of the embed-fuse cycle, and how rules exceed what ordinarily goes on in symbolic calculating, to violate invariant theorems and collective norms. In fact, there are never too many of these examples. Every time I'm sure people understand one, they show me they don't—the embed-fuse cycle is a simple idea that seems hard to keep. But some of my examples are tricky, almost magic—mysteries couched in the fickle eye. Three of my favorite examples provide the locus for this exhibit; in the trio, rules rely on embedding to exploit the inherent ambiguity in shapes that fuse. There are many surprises, all packed with boundless delight.

**Example 1.**

Let's start with three equilateral triangles in a pinwheel

and rotate the pinwheel about its center

This is easy to do in the obvious way with the rule

**Exhibit 2**                                                                 79

in the transformation schema x → t(x), that rotates a triangle about a vertex, pinned at the center of the pinwheel. I can use copies of the rule three times in the schema x → Σ F(prt(x)). Then, x is the pinwheel and prt(x) is each of its three triangles, so that different versions of F change (rotate) these parts separately in a coordinated sum. This describes the twisting action of the entire pinwheel in the rule

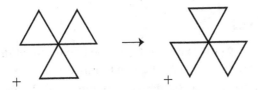

My description of this rule puts a pin at the center of the pinwheel to fix it in place as I rotate it, but there are alternative ways to track this movement in the rule. In fact, each of the triangles might have its own independent path, or divide into parts that move separately. Just looking at the rule, without knowing that it's the sum of three triangles and their rotations, I can't tell what motions are actually involved. Any will do that begin and end in the right way. Maybe the entire pinwheel takes a path like so—

The two sides of the rule limit the transformation at its start and at its finish, without showing what happens going from one to the other. I can do this with other rules, usually in x → t(x), for the pinwheel and its parts in any way I please to spell out their motion in ever finer detail—inbetweening in a kind of animation. Now suppose I want to use copies of the rule

$$\triangle \rightarrow \triangledown$$
$$+ \qquad +$$

for F in the schema x → Σ F(prt(x)), to get the same result. This seems pretty unlikely, even spooky black magic. I guess strange stuff goes for more than particles and action at a distance—my new rule rotates a triangle about its center that's a fixed point, but the centers of triangles still rotate just as the pinwheel does

How can a rule that keeps the center of a triangle fixed/pinned in one place move it at the same time to another place? (The fixed/pinned points given explicitly in this case and in the previous one may be viewed as 0-dimensional symbols with associated parameters for triangles. The trick is to trace a way that shows how higher-dimensional elements in shapes—here, 1-dimensional lines—can be used to go around symbols in descriptions/representations.[1] Then embedding isn't identity, and brings in art and design.) I said it was spooky; in fact, it sounds totally absurd. Is this a new kind of weird paradox? Let's see what happens as I try it out. I can calculate just so—

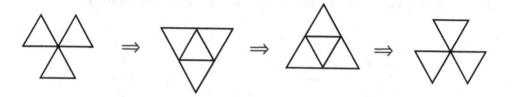

In the first and third steps, the rule

is used three times in the schema $x \to \Sigma F(prt(x))$, and in the second step, it's used twice. The shape

**Exhibit 2**                                                                                      81

is produced from three triangles and changed as two, while the shape

is produced from two triangles and changed as three. In the transition between these two shapes, the centers of the three triangles in the starting pinwheel are moved implicitly to the centers of the three triangles in its rotation—more precisely, when the large, outside triangle is rotated about its center

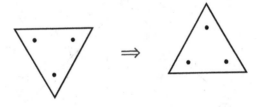

or reciprocally when the small, inside triangle is

The three-step process from pinwheel to pinwheel is embed-fuse, embed-fuse, embed-fuse. This goes smoothly without a break, because the triangles in the second and third shapes fuse. As a result, the shapes are ambiguous with respect to the rule

In retrospect, the rule shows how this works in topologies for the shapes

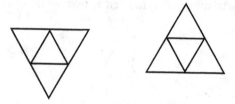

and the limiting pinwheels, to ensure continuity as I calculate.[2] But these relationships may be better summarized in another way—in this graph

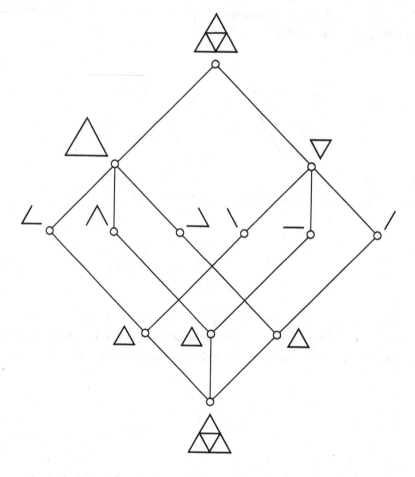

to reconcile incompatible descriptions. The top shape is two triangles in a tree defined using the identity for triangles

**Exhibit 2**                                                                 83

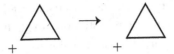

in the schema x → Σ F(prt(x)), and the bottom shape is three triangles, in terms of the same identity. Going from top to bottom in the graph shows how to combine the parts (units) at the six vertices of the middle tier, in order to change two triangles into three; going the other way around, three triangles are changed into two, combining the parts at the middle tier in a different way. The parts fixed in the middle tier are the products of the triangles/parts distinguished in the trees at the top and bottom of the graph—these latent (emergent/emanant) parts are required to make sense of what I see in alternative ways, in non-hierarchical, contradictory/contrasting and incompatible/inconsistent views. The following table shows how these products are defined—

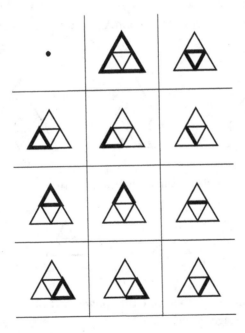

Sometimes, in a bit of cerebral legerdemain, I tell myself that this kind of table is a nifty "symbol machine" that augments the ABC's. The parts it produces trace new symbols to label/name the nodes of the graph—to highlight the shift from shape to structure, etc. It may not feel that way, but seeing is a lot more complicated than it looks at first blush, especially when definite parts are involved. That's why schemas and rules for visual calculating in shape grammars are so valuable; they let me calculate directly in terms of what I see, with no intervening structure to get in the way. There's no reason to worry about underlying parts in visual analogies, or merging conflicting trees in graphs or defining topologies to ensure continuity for rules. Visual analogies and the like, given before I calculate, only make things hard; they're defined meaningfully only after the fact, as a result of calculating—not before rules are tried, not before I've had a real chance to see as I please. My example also suggests others in a host of different ways, maybe first off, in the series

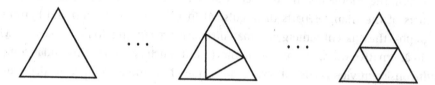

in which shapes vary parametrically as John McCarthy and Franz Reuleaux urge, or in the series

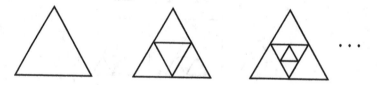

in which triangles are inscribed recursively using the rule

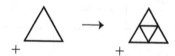

in the addition schema x → x + t(x). Other polygons work for all of this, as well—squares, pentagons, hexagons, etc.

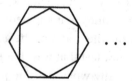

Exhibit 2                                                                          85

in an ongoing joust between pairs of polygons and multiple triangles. And irregular polygons are good, too—easy variations are everywhere I look. Perception is fluid, always in flux. There's no way around the embed-fuse cycle—it's the lifeblood of visual calculating.

## Example 2.

Let's try my second example; it's a problem in object-oriented programming that was initially noticed a little more than two decades after it was already solved algorithmically, using shape grammars. (The original solution I have in mind is in *Pictorial and Formal Aspects of Shape and Shape Grammars* in terms of pictorial equivalence, subshapes, and reduction rules. A few read it now—more should.[3]) This bolsters my long-held view that art and design have a lot to contribute to computer science, or at least to calculating. Tellingly, the computer relationship doesn't seem nearly as strong going the other way around—so much for the wild claims and tiresome hype. But wait and see what my example holds. In his book *On the Origin of Objects*, Brian Cantwell Smith considers computers and intentionality. Sometimes in my weaker moments, I worry about the intentionality of his title that oddly nods to biology and Darwinian evolution.[4] Nonetheless, Smith frames the issues nicely, when he "presses for a kind of representational flexibility that . . . object-oriented [computer] systems lack." This is evident in BIM and parametric modeling, and whatever comes up in computer-aided design (CAD). (For more on computers and AI in this vein, see note 11 in this exhibit.) In fact, representational flexibility is key, and not just for object-oriented systems. Surely, it's vital for poetry in the sweep of metaphor and all kinds of figuration ("a bush supposed a bear"), and then for Coleridge's magical esemplastic power (imagination), when things fuse in order to re-create (re-divide)—"all objects (*as* objects) are essentially fixed and dead." Smith shows why this kind of representational flexibility is elusive for computers, using an entirely visual example—

> Many current systems are not only remarkably inflexible, but tend to hang on to ontological commitments more than is necessary. Thus consider the sequence of drawings shown [just below]. Suppose that the figure shown in step 2 was created . . . by first drawing a square, then duplicating it, as suggested in step 1, and then placing the second square so as to superimpose its left edge on the right edge of the first one. If you or I were to draw this, we could then coherently say: now let us take out the middle vertical line, and leave a rectangle with a 2:1 aspect ratio, as suggested in step 3. But only recently have we begun to know how to build systems that support these kinds of multiple perspectives on a single situation (even multiple perspectives of much the same kind, let alone perspectives in different, or even incommensurable, conceptual schemes). It should not be inferred from any of these comments [however]

that these theoretical inadequacies have stood in the way of progress. Computer science does what it always does in the face of such difficulties: it makes up [ad hoc] answers as it goes along [debugging after the fact]—inventive, specific, and pragmatic, even if not necessarily well explicated.[5]

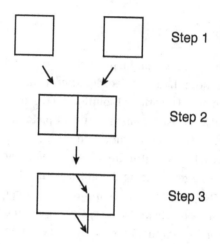

A big sigh—what a Sisyphean task. I guess explication isn't necessary. There's scant reason to understand ambiguity for good and all when you can patch things up time and again as you need to, with cunning, after the fact as you go along.[6] (Of course, this is ironic for computers that try to predict what you're going to see.) Too bad there's so much to fix. In fact, endlessly more, and more that's likely to clash or interfere with what you've already done—patches on patches kept out of sight, in perpetuity. (The tragedy of false/fake answers is abundantly clear in Example 3.) Just what makes seeing so rigid for computers, when it seems so fluid to us? Maybe this is what Sisyphus really shows, that hidden cunning (debugging) is simply fruitless toil. Many try visual analogies (conceptual schemes) to cheat seeing, just as Sisyphus cheats death in the myth; then what next?—to start over, rushing frantically to put in what's been left out, or to give up and insist on a single, "coherent" response that must be rigorously taught in rote recitation, and endlessly tested to make sure that everyone gets it right and vigilantly policed to keep it so. Either way, there's no end in sight. Or maybe this is a comprehensive plan for job growth and job security. And I thought computers were supposed to be labor-saving devices, the marvels of modern technology. Yes, to see things in new ways, to go on easily without breaks or gaps to bridge, is an intractable problem—a challenging conundrum that begs for a single solution that seems to be forever out of reach, when it's tried piecemeal in descriptions, part by part in visual analogies or equally,

**Exhibit 2**                                                                                    87

symbol by symbol. Somehow nothing seems to add up in the desired way—so that one shape is many and many shapes are one. And such problems arise time and again today in computer science and AI, in surprising places when insight matters more than usual, if reasoning and logic are going to work—for example, in calculating the combined area of a square and an intersecting parallelogram beyond its sides

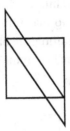

that are also two intersecting triangles. Maybe there's a straightforward way to find an answer that doesn't wait for difficulties (blind spots) to arise and multiply/compound with every change in perspective; maybe there are real opportunities in this. In a shape grammar, I can define the two squares in step 1 of Smith's little example using the rule

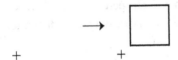

in the inverse x → prt⁻¹(x) of the schema x → prt(x) for parts, and the rule

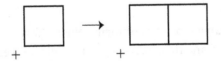

in the addition schema x → x + t(x), to duplicate x in any way I choose, anywhere I want it to be, using any given transformation t. In x → prt⁻¹(x), x is the empty shape (here, a blank space), and prt⁻¹(x) is a square (in fact, prt⁻¹(x) can be any shape, because the empty shape is part of every shape); in x → x + t(x), x is a square, and t is typically a garden variety reflection or a translation that's the side of the square. This produces the shape in step 2 without any fuss. Then the identity

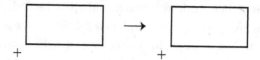

is enough to find the rectangle in step 3—it simply embeds the rectangle after the two squares fuse. In fact, the rectangle and line are defined explicitly, when reduction rules are applied to lines (basic elements) to make the maximal (longest) lines that correspond to what a person trained in drafting the old-fashioned way would draw with a T-square and triangles. This is the smallest number of lines that describes the shape, a kind of visual Taylorism for drawing—

This highlights the difference between combining two squares as objects (symbols), and combining two squares as shapes. Objects combine/merge as elements in sets (visual analogies and spatial relations) or in other kinds of set-like structures, and they're preserved separately and are never tampered with; whatever parts there are, are simply different combinations (subsets) of distinct and independent objects. Two squares have only these parts

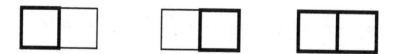

none of which are rectangles or parts of them. When shapes combine in shape grammars, however, they fuse, and then, parts depend on rules and embedding. Two squares aren't defined explicitly in these five lines

But the squares are there "figuratively," because I can embed their sides in the lines—try it with tracing paper in a visual proof to see how lines and segments match up. The shape

**Exhibit 2**                                                                89

isn't just two horizontals and three verticals, or a rectangle and a line, or anything else I can say—I can see in it whatever I choose to see, whatever I can trace in its lines and their segments, whenever I try a rule. It's worth trying this to see what you get; it's always more than anyone ever assumes in advance, much more. No good counting in your head—it's what you see as you draw. Suppose I trace everything inside a simple curve—

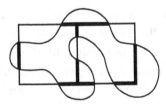

This may be a practical method to design building plans, complete with walls and rooms, windows and doors, etc. Designing isn't simply combining things in a lot of different ways and sifting through the results to find what you want; it requires seeing surprising things in other things, in the ones you're designing and in the ones you've copied—"in a tree-trunk or clod of earth," in John von Neumann's Rorschach test or Oscar Wilde's beautiful form, in prior designs you're sure you know/understand that alter freely nonetheless. All of this takes insight and imagination, with embedding and shapes that fuse. Or I can try the rule

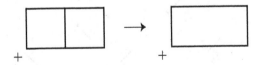

in the schema x → prt(x) to erase the middle vertical line to make a rectangle. In this case, there are two obvious steps; calculating goes like so—

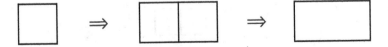

Once again, this relies entirely on the embed-fuse cycle—it's impossible to do without it. Two squares fuse, so that I can put in a rectangle or find other parts of any kind I like. Moreover, the alternative ways of seeing the shape

as side-by-side squares or a rectangle and a line can be elaborated in a graph to merge distinct ("incommensurable" or at least topologically incomparable) trees in the same way I used an identity in Example 1 to show how two triangles are actually three, and vice versa. (This kind of incommensurability and topological incomparability go for the trees in Example 1, as well.) But surely, squares aren't a rectangle and a line. Isn't that why computer systems in art and design (BIM, etc.) are so successful? They keep everything apart, never coadunate. Squares, rectangles, and lines are individual objects of three separate kinds/types that don't interact; at least they shouldn't for things to make any sense—

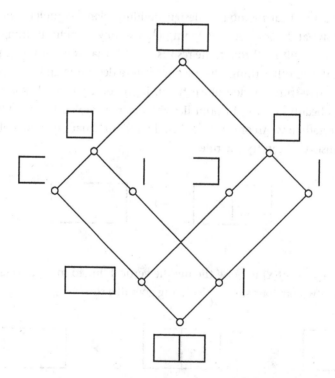

**Exhibit 2**                                                              91

In my new graph, squares, rectangles, and lines are interchangeable. The identity

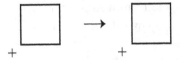

for squares defines the tree at the top in the schema $x \rightarrow \Sigma \, F(prt(x))$. As required, the squares overlap—the right edge of the left one is the left edge of the right one. In the same way, the identities

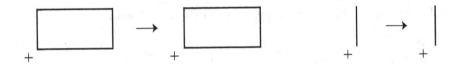

pick out the discrete rectangle and line. The superimposed (right and left) edges of the squares are represented as two lines in the single line that divides the rectangle. This is easy to see in the table of products for the parts picked out in the two trees—

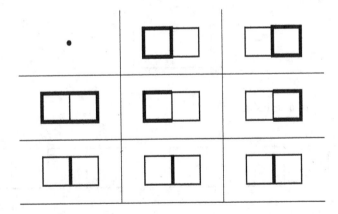

Of course, what's coherent isn't anything anyone can say, or coherently say, for sure—there are untold ways to see the shape

At the very least, there are building plans galore (indefinitely many of them), traced with the identity for lines—but really, the identities for squares and rectangles are a better start. They do a lot more than Smith suggests, maybe more than steady reason and thought are ready to allow. Side-by-side squares are equally a rectangle and a square, and a rectangle and two squares—

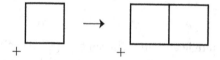

Neat, but who would have ever guessed? And why not use the rule

recursively? The shape

is merely the first in a series of shapes

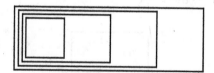

made up of dilating rectangles in arrangements like this

**Exhibit 2**

93

and in many other arrangements, as well—

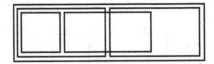

(It's worth it to spend a little extra time on the schema x → Σ F(prt(x)), and the inverse of the rule

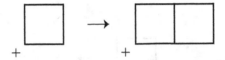

that is to say, the rule

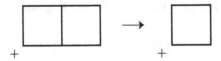

that erases one of two adjacent squares. I can use the inverse rule to insert gaps and spaces in rows of squares—for example, like this

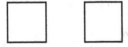

when x is the shape

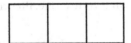

and prt(x) is the pair of left squares and the pair of right ones. The inverse rule is applied to both ends at the same time, so that neither of the end squares has an interior side that's erased—

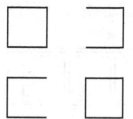

Clever, although the same thing happens elsewhere, sometimes with disappointing results—for example, the part schema x → prt(x) includes the erasing rule for squares

that takes out the middle square directly in the obvious way, to produce the shape

in which there are no squares at all. How can I ever fix that? Also for the pinwheel in Example 1, there are unwelcome surprises, when I try the rule

to rotate each of its triangles about a vertex, one after the other in the sequence

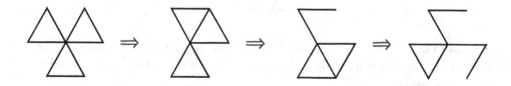

**Exhibit 2**                                                                 95

And I can go on with the last triangle, to end with it alone. If squares and triangles were objects/symbols, this would never come up—it wouldn't be a problem. Is there a way to keep parts the same as needed, without using symbols that are invariant forever? In fact, the identities x → x fill the bill. They do an impeccable job of reconstruction and repair—simply put in the identity for anything you want to save, or that's to stay the same, with everything else you've already combined in x → Σ F(prt(x)). Try it with the inverse rule

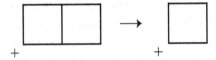

and the familiar identity for squares

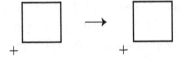

for one end the shape

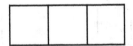

or the sum of the identities for the squares at both ends—

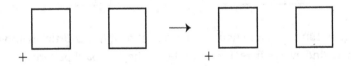

And the identity for squares goes equally with the erasing rule for squares

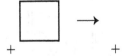

For the middle square in a row of three, the rule with everything in a sum is

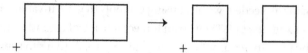

And playing around with erasing and identity in longer rows of squares, maybe staggered ones, shows more. It's a neat trick that works anywhere you please—because identities are incorporated fully in the rule that's defined in $x \rightarrow \Sigma \, F(prt(x))$ before the rule is tried, changes and corrections occur together in a single process that's integrated from the start. Sometimes, surprises are welcome, and sometimes, they're really not. Identities work for parsing, and now as well, to ensure that given/selected parts are invariant, at least for the time being. Identities keep parts intact, offsetting any unwanted side effects that may result as other rules are put into use—and the side effects needn't be known in advance, only what's to be saved.[7] Of course, other kinds of relationships for identities and rules are also possible to describe, parse, and produce shapes and pictures, using constructive and evocative devices in which perception and meaning interact and vary freely.[8] This adds to the algorithmic approach to aesthetics that I outlined in note 29 of "Seven Questions." It's pretty amazing how much identities do.[9]) The way rows of squares are rectangles extends nicely to grids and similar forms

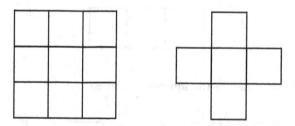

It's always great fun to explore figure-ground relationships, effortlessly in varied grids; they're easy to find everywhere, for example, in checkerboard patterns

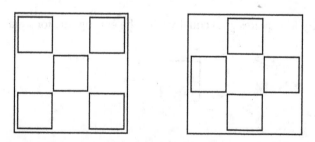

Exhibit 2                                                                                                        97

so that presto chango, five squares are four. Then, there's the coloring book schema form "Seven Questions" $x \rightarrow x + b^{-1}(x)$ and the more inclusive schema $x \rightarrow x + b^{-1}(prt(x))$ to fill in boundaries—

There's no end to schemas and rules. Visual calculating in shape grammars goes beyond what symbolic calculating allows in computers with visual analogies and spatial relations—conspicuously in object-oriented systems in CAD and BIM. Surely, "representational flexibility" is a praiseworthy goal, but a handful of coherent descriptions, whether Smith's or yours or mine, is bound to be sorely incomplete when it comes to a Rorschach test or a beautiful form. Embedding and shapes that fuse ask for much more—plenty for everyone, and then extravagant descriptions (prodigal ones) to try at will. There's no telling what any of us will see next time that exceeds everything we've seen before, that's indubitably coherent and makes perfectly good sense now (in original participation) and retrospectively (in final participation). This kind of personal freedom (representational flexibility) to see as you please is at once intuitive and open-ended—multiple perspectives abound—but it seems that there's never enough freedom in computers for art and design, or for very much of anything else when it comes to shapes and rules. In order to have enough, there must be too much. In shape grammars, too much is never less than all there is—the embed-fuse cycle brooks no limits.[10] Seeing is relentless; the winged eye never touches ground.

## Example 3.

My third example runs straight to the goal—it compares shapes and symbols directly, and shows how they differ. In fact, it comes from an earlier essay of mine on Wilde's critical spirit, with its key aesthetic formula to see things as in themselves they really are not—"*The Critic as Artist*: Oscar Wilde's Prolegomena to Shape Grammars."[11] I've modified this in a few places, so that it fits easily with "Seven Questions" and adds more to it, but in spirit it's what I wrote before. In computer science (actually, in the theory of formal languages and automata), a rule in a generative grammar (Turing machine) is useless if it contains a "nongenerating" symbol or symmetrically, an "unreachable" one in a given vocabulary of symbols.[12] Unreachable symbols provide another way to show how shapes and rules work in shape grammars, and why shapes aren't symbols. In outline, the core idea is this—a symbol is reachable when it's in the initial string of a grammar or recursively, in the right-hand side of a rule that has only reachable symbols in its left-hand side. Rules with unreachable symbols can't be used to calculate; there's no way to apply them to any string I can get from the initial string using the other rules in the grammar, because every such string contains reachable symbols only. This doesn't seem very hard to check—maybe my initial string is the same symbol S lined up three times in this concatenation

SSS

and my rules are these five

S → AB

C → BD

AD → ABC

SA → a

B → b

Starting with S as the first reachable symbol, and working recursively, I get

S, A, B

because A and B are in the right-hand side of the rule S → AB, that already has a reachable symbol in its left-hand side, and then I finish with

S, A, B, a, b

because a and b are in the right-hand sides of the rules SA → a and B → b, with the reachable symbols S, A, and B in their left-hand sides. The five reachable symbols are S, A, B, a, b, and the unreachable ones are C and D. So, C → BD and AD → ABC are useless rules that I'm free to delete—the grammar works the same when they aren't there. This is unremarkable for unreachable symbols; there's no way to try the rules in which they occur. But is this the case when symbols are shapes? Can I tell whether shapes are unreachable or not? Let's see how this unfolds if I try to discard useless rules in shape grammars. With shapes, there are bound to be some surprises. Suppose I start out with an initial shape made up of three squares, that corresponds to the initial string SSS with its three S's in a row. Maybe the squares are arranged, haphazardly or not, in a constellation like so—

**Exhibit 2**                                                                99

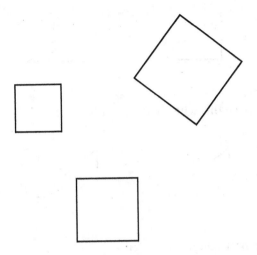

And suppose I calculate as usual in a shape grammar with the rule

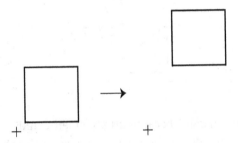

in the transformation schema x → t(x), that translates a square back and forth—

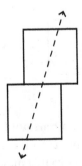

Of course, I can add other rules, too—maybe this one

in x → t(x), and also from this schema, the inverse

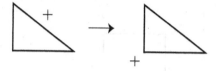

to translate a triangle back and forth, as well—

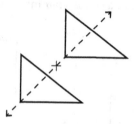

I need two rules for the triangle, because it's asymmetric, unlike the square. But does adding these two new rules do any good? The triangle in their left-hand sides isn't in the initial shape—obviously there are exactly three squares. I can use the erasing rule for squares

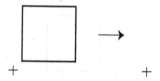

to make sure, eliminating them one at a time until nothing's left—isn't this the proper way to count things out, object by object? Nor is the triangle located at any place I'm able to see in the right-hand side of the rule that moves a square. Everyone agrees that triangles aren't squares—it's no big deal, the triangle is definitely unreachable, and the rules

**Exhibit 2**                    101

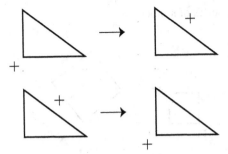

are undeniably useless. I guess my rules for triangles are a mistake. But let's see if it's prudent to say this, that it matches my ongoing visual experience when I calculate. If I try the rule

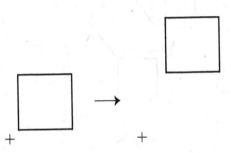

on the largest square and on the smallest square in the initial shape

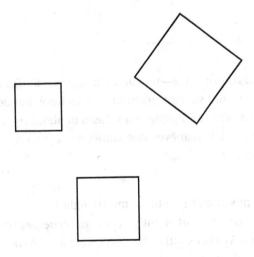

in that order, I get this series of shapes

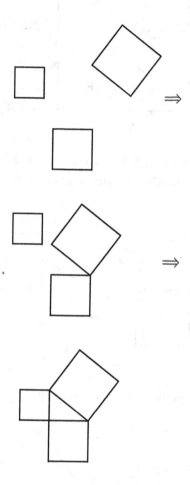

that ends with a Pythagorean figure—that unmistakable symbol of number, rationality, and usefulness. What would students learn in grade school and beyond, especially in STEM subjects, if there were no Pythagorean theorem about right triangles and their sides? It's the eternal formula that everyone knows by heart and can recite effortlessly. It's easy to say in words—

$a^2 + b^2 = c^2$

But this is merely symbols and numbers—my triangle is $3^2 + 4^2 = 5^2$. What made it appear as if by magic, when I started out with three squares, as counting proves? And in fact, God in Norbert Wiener's little, children's song in *Cybernetics* would agree that counting is the hallmark of definite things, so the squares must be there.[13] I guess the

**Exhibit 2**                                                                 103

squares are arranged around the triangle—that's easy enough to see—but this isn't a binding distinction in shape grammars. I can embed (trace) the triangle in the Pythagorean figure, so it's there, too. No matter how I try to finesse this, I can apply the rule

to translate the triangle and thereby calculate some more to get

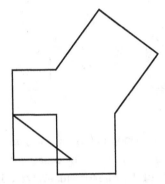

and this hardly seems useless. Just the opposite, it might be very useful. (I guess squares aren't objects in shape grammars, like the two squares in Example 2 that form a rectangle that can't be defined.) And I can go on in this manner to move the triangle again

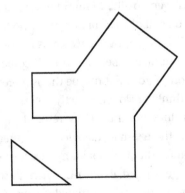

That's kind of strange—who would ever have imagined such a thing could happen? Now my first rule for squares

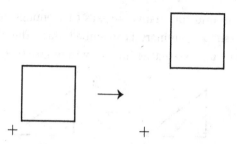

the one that was so useful before, is useless. I guess this goes for symbols, too—useful rules can become useless, but then it must be forever. In my figure with triangles, I can use the inverse

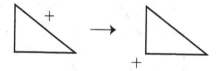

twice to retrieve three squares that pop out all at once, and then move them separately in multiple ways—eight ways for each one to be exact, in terms of how the symmetry of the square in the left-hand side of the rule for squares is modified by its right-hand side.[14] Useless/useful rules can turn into useful/useless ones, and then switch back again—in untold ways. (Surely, this is a marvelous source of original plots worthy of Wilde.) A trio of useless rules—two with an unreachable triangle and one with an unreachable square—that I thought I could discard without any qualms, are perfectly useful in obvious ways but at different times. Is this even possible? I must be missing something important—but really, visual calculating and symbolic calculating can't be that different. Calculating is calculating, isn't it—just as a rose is a rose? This seems reasonable—everyone takes it for granted—but it's probably a mistake, especially when it comes to what I see, and maybe what I hear when I say "a rose" (eros). But maybe this is merely a misunderstanding that can be resolved rationally, thinking slow in strictly logical steps.[15] Suppose that squares and triangles are really four lines and three lines, respectively, as I was taught time after time in school. Why not use these visual analogies (spatial relations) in place of shapes? There must be a good reason for them, although my teachers were totally baffled when I asked for one—"why is George worried about this, when it's so obvious?" But I guess my teachers actually knew best—if squares and triangles have distinct sides, then the side of one can be the side of the other at the same time, in a kind of mutual exchange that defines both. It seems that lines can be like symbols (0-dimensional elements) in visual

**Exhibit 2**                                                          105

analogies for squares and triangles, and work just like units as I calculate—in effect to make visual calculating the same as symbolic calculating. Being practical works wonders, taking things as in themselves they really are, at least the way they are, as they're rigorously taught in school—education is an admirable/necessary thing in an impossibly ambiguous world. This highlights the key difference between set grammars for spatial relations and shape grammars.[16] I've already tried shape grammars, but this merely confused useful rules and useless ones. Maybe set grammars will work better, to avoid making the same mistake again. The initial shape, exploded a little to separate symbols, is now 12 lines

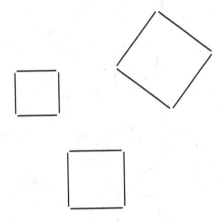

that are all units, and my three rules, also in exploded versions, are

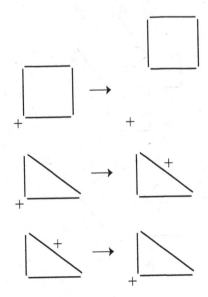

Lines are reachable—they're in my initial shape and the only symbols in my rules—and so, no rule is useless, even if it's hard to imagine how rules for triangles might be used. That's the reason to calculate—to find out. And when I do, everything works exactly right—

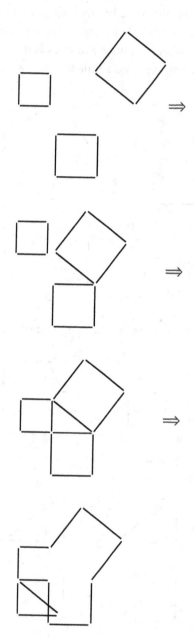

**Exhibit 2**                                                                                                    107

This is kind of cool. But are these visual analogies really a sure thing? They seem too good to be true. Does everything always come out right if I keep strictly to the lines that define squares and triangles? What can go wrong? Do I need to add more detail to be positive that these visual analogies are complete? How and when do I know this once and for all? Are there examples that don't work, even in theory?[17] Consider the little triangles—the so-called "emergent" ones

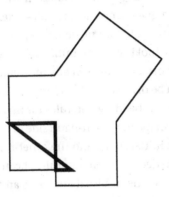

I should be able to move these latent parts with rules. And it's easy in shape grammars with embedding, ironically because triangles are unanalyzed and undivided, and not lines (symbols) in a visual analogy—so much for thinking slow, with what I learned in school. And how about the emergent square

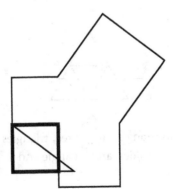

In fact with these surprises, my three rules with lines should all be useful at the same time. That's something to be happy about. But I can't move either of the triangles or the square because they don't have sides—lines (symbols) put in by my rules—or they

have fewer sides than they need. And suppose they did have sides, then what would happen to the lines in my three initial squares and in the emergent triangle they contain? How many additional rules are part of the mix to make any of this work? How many definitions of squares and triangles are there? All of a sudden, things seem to be getting awfully complicated—simply trying to move squares and triangles around in an effortless way. And what all of this shows is how truly different shapes and symbols actually are. Neither individual shapes nor shapes in rules can be described before I calculate, while symbols are given once and for all—fixed and independent forever. It may be that squares and triangles are made up of lines, but how do I know which ones? Can I find out just by looking at my initial shape and my rules? This is OK in generative grammars—symbols are known in advance. But it seems hopeless in shape grammars, because lines can be divided freely in lots of ways. I can't decide what I'll see before I've seen it, and I need embedding and rules for this. I have to calculate in shape grammars, to find out if the shape in the left-hand side of a rule is reachable or not—it's never enough to look at the initial shape and the other rules—and this makes it hard, no, it makes it impossible, to decide. There's nothing I can do before I try my rules, to find out what's going to happen next. Visual analogies and other descriptions aren't a reliable way to start. They rarely work and are usually misleading without calculating first, without them. And then, what kind of calculating is this—surely, it's not calculating with symbols? There are many possibilities that are worth knowing—but for most of them, this means going on, not knowing beforehand they're there. There's nothing to learn in school that helps, not even recognized things like squares and triangles. For example, for the shape

in Example 1 that's both two triangles and three, triangles are described in three distinct ways, as three angles, three sides, and an angle and a side. Go figure—

**Exhibit 2**                                                                                      109

I guess that's finally it for my teachers—I had good reason to worry. It's exactly as Wilde puts it—"Education is an admirable thing, but it is well to remember from time to time that nothing that is worth knowing can be taught."[18] That's because what's worth knowing and what I value changes as I calculate, in a continuous process that's full of surprises. The use of embedding for shapes that fuse makes a huge difference—useless rules turn out to be useful. And maybe that's it for uselessness, too—it's not a distinction that makes much sense in shape grammars. It's a cinch to cook up more examples like mine for squares and triangles, for any polygons—and equally in fact, for any shapes made up of linear elements.[19] Any shape or vocabulary of shapes can be used to make other shapes, sometimes in surprising ways. For example, three squares, a small one and two larger ones of the same size, combine in this shape

The rule is

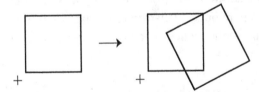

in the addition schema $x \rightarrow x + t(x)$, and it applies twice to a single square. And it's pretty clear that the lines in an A are embedded in the sides of the three squares

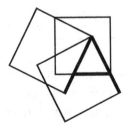

or is it two A's that are embedded in the squares

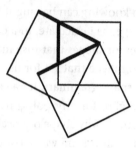

Moreover, the distinct symbols A and E contain the expression A + E in a similar way, no matter that neither A nor E is/includes +.[20] See for yourself—

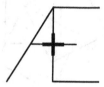

Symbols are shapes, too—or have an external presentation (sound, surface texture, etc.) that can be sensed/perceived—and as such, they're something to embed in shapes, or find in like phenomena. An A isn't just in three squares, but in two or more in many arrangements. And A pops out elsewhere in neat ways, for example, first in the remarks of an important philosopher, when he adds a triangle and a hexagon together, and wonders what this means—

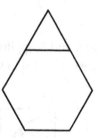

Then fancifully, it's in Johann David Steingruber's architectural ABC.[21] Steingruber puts rooms and walls together, with doors and windows, in elaborate building plans to make all the letters of the alphabet, here in an extraordinary "SECOND A"—

**Exhibit 2**                                                                                                 111

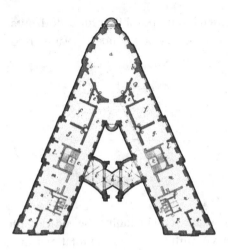

Which description of this shape fits it to a T, as in itself it really is—or is not? A building plan replete with rooms and walls, etc. represented as data/objects in BIM, the letter A, half a raven, right/left leg and partial beak (memory is lost when rules are tried, but it may matter in other ways), an upside down antelope's head, a philosophical puzzle, any combination of the five descriptions, none of them, or still additional ones? And once Steingruber's A is seen, it's impossible not to embed a copy (transformation) of it

as the third A in my three squares. An identity does the trick, to put in (find) the center of this symmetric trio, with overlapping A's that are lined up as you please, maybe staggered and spaced in elevation, front to back, left to right, whatever you want—

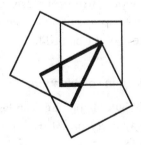

This copy, though, is merely one possibility for a horizontal hinge that opens and closes, à la McCarthy and Reuleaux. Steingruber's A varies in parametric profusion from greater than 0 to less than $2\pi$—

θ = 0            θ = π/2                  θ = π                    θ = 3π/2            θ = 2π

(In this series, I can change any term including the verticals into any of its successors in the schema $x \rightarrow \Sigma\ F(prt(x))$, with two reciprocal rotations and a reflection in $x \rightarrow t(x)$. The term and its successor are divided with respect to x and t(x), each into four component parts/lines that are equal from term to term.[22] As the series extends symmetrically past $2\pi$, the hinge folds down to the base of the A, instead of up to its top. This kind of switch also goes for Reuleaux's series of nuts and bolts.) It's uncanny—how the two kinds of A's in my three squares are the same in one series. There are surprises galore whenever I open my eyes, and remarkable things to see—for example, two exaggerated A's in my three squares that aren't in my series and maybe a third, and the smaller M in Steingruber's A that's many other things, but this is enough for the time being. It seems that visual calculating in shape grammars isn't the same as symbolic calculating in Turing machines. In Turing machines, symbols are simply as in themselves they really are. Isn't this invariance what makes calculating possible in the first place—every string of symbols is easy to read as an elaborate sort of coded number. What would happen without invariant symbols? What if they weren't always the same? What if symbols were each and every one a Rorschach test? I guess this spells the end of Wilde's aesthetic formula with its warrant for ambiguity and unbridled change, and of pictures and poems, and art and design. It appears that calculating isn't seeing—unless of course, calculating is as in itself it really is not. How can this possibly be true? But that's what I've been trying to prove in "Seven Questions" with Coleridge, and von Neumann and Wilde, and in my expansive notes that tie in lots of others, and now, in this exhibit with my three examples to show the sweep of the embed-fuse cycle. Maybe there's a way to put all of this together in a single sentence—

Shapes aren't symbols, alone or in combination.

This is the quick of art and design. Shapes are seamless—they're divided again and again on the fly, in another way every time I try a rule in the embed-fuse cycle, to

Exhibit 2                                                                113

calculate with what I see. The ambiguity is what makes intuitive/personal experience possible, for example, in Wilde's modes of reverie and mood with countless changes and contradictions, and limitless delight.[23] Seeing—trying rules in shape grammars—goes on in ever shifting ways, that no one dares assay in advance. There's no abstraction/structure, no chart, no map, no survey that works. My Pythagorean shapes, with squares and triangles that are reachable and not in a single shape grammar are, in fact, entirely commonplace, but they're unexpected just the same, even to the well-traveled eye—

> The three squares—just floating down into the Pythagorean figure. And then those two little triangles, and the fourth [emergent/latent] square. The first time you see the square and triangles, it's a total surprise. Once you've seen them and look again, there's an inevitability, a destiny to it. A Greek tragedy for calculating with symbols. The inevitable outcome of its tragic flaws.[24]

The Greek spirit is Wilde's original source for the critic/artist, who's armed with an aesthetic formula for different descriptions—Epicurean ones that are all true. I like this description of the Pythagorean figure because it contrasts shapes and symbols as the former lead to surprises that overwhelm the latter. And surely within Wilde's critical compass, reverie and mood go anywhere that embedding desires. Unexpected or not, tragedy is inescapable as long as symbols describe shapes fully, and once and for all in terms of what's previously known. No matter how much effort is put into this, shapes exceed any visual analogy (description) that's fixed in advance. Shapes are ineffably sublime—there's no saying what they really are as long as there's another rule to try in the embed-fuse cycle. But whenever I talk about how shapes and shape grammars are related to art and design in this way, someone invariably feels compelled to explain to all and sundry everything that I've misunderstood—and isn't this as it should be, isn't misunderstanding inevitable, being the starting point for seeing things as in themselves they really are not? Sometimes, my interlocutors have timeless lessons to teach about how things really are in pictures and poems, and sometimes, they betray sincere regret, as if I'm desperately lost in a kind of personal (psychotic) reverie that knows no bounds. Aficionados of the arts sigh and shake their heads—he just doesn't get it. But they can't say how or why, because then I could try and might succeed—or maybe not. Scientists and engineers aren't that much better, and they may be worse—that's not calculating, because you can't put it on a computer; you're just gaming the system; who told you, you could calculate like that? Science has some curious conventions that must be kept, but I prefer to see for myself. No matter who stands up, they're keen to recite tried and true slogans—McCarthy's is a good example—that lend comfort and support to the artist and the scientist alike, so that each can dismiss whatever methods and devices the other finds so dear. There's a thesis and an antithesis, and no synthesis. Yes, you can wear two separate hats, each at a different time; you can be a painter/poet on a Monday and then a physicist/mathematician on a Tuesday. But this frustrates and may even bar imagination to ensure two distinct cultures in one common world—art and science, and their vast unknown interstices never fuse in order to re-create, in unexpected divisions and precincts.[25] New insight is invariably lost in the incessant din of a rote chorus that repeats, yada-yada-yada, the kinds of things that everyone takes for granted—

Art is breaking the rules.

No doubt, this aesthetic rule is included in Wilde's aesthetic formula. But somehow the rule ensures that art and calculating are irredeemably apart, even if the formula is good for both. It seems that art breaks the rules, so it can't possibly be calculating, because calculating can only follow the rules—for computers, it's drag and drop, click-click, click, click, click. Maybe so for symbolic calculating and unreachable symbols, but this isn't so for visual calculating in shape grammars, where breaking a rule simply means trying another one, or using familiar rules in alternative (surprising) ways in the embed-fuse cycle. That's why Wilde's critical spirit provides such an invaluable prolegomena to shape grammars. The critic as artist invokes Wilde's aesthetic formula to see things as in themselves they really are not—every time with another rule. I'm free to see in this way, as well, in any way I want. I can draw two squares

and then see four triangles, or pentagons, hexagons, K's and k's—and in fact, right angle A's. There's no end to breaking the rules in shape grammars and visual calculating. Ambiguity and like uncertainties—change, contradiction, discontinuity, incompleteness, inconsistency, etc.—are the nub. Some of my protagonists in "Seven Questions" promote this, and some don't. But getting any of them to agree isn't the problem—I can keep them together with a little negative capability. Rather, the trick is to exploit ambiguity, as I calculate. This requires embedding and shapes that fuse, so that any rule works, no matter when I try it. Whether I'm consistent/coherent or not doesn't make even the slightest bit of difference—I'm free to go on. Nothing blocks my way. It's

**Exhibit 2**                                                                                 115

worth repeating—rules apply with no inherent memory, so I'm seeing for the first time, every time I look. And what I see—anything at all—corresponds to a rule. There's no way around this, because what I see relies on the embed-fuse cycle. Of course, the insistent question remains of whether I can predict the next rule I'll try or not. In answering this query, I invariably opt for not—picking a rule to use is exactly the same as putting an interpretation into a Rorschach test. Each of us calculates in a unique way—in von Neumann's words, as "a function of [our] whole personality and [our] whole previous history." But then, von Neumann's Rorschach test is Wilde's beautiful form—both are engaged personally with rules. It seems pretty sure that artist-computers (visual analogies) aren't enough, without the embed-fuse cycle in shape grammars. The rule I try, whatever it is, works now, whatever I see—and there's no reason to tailor any rule for this in advance. Nothing is left out to cheat seeing. Prior descriptions and tedious debugging aren't required to go on—Sisyphus is free at last. Rules forget what I've seen and may neglect any/all of my well-made plans; they needn't respect the past to work, or anticipate and plan for what's to come. Embedding and shapes that fuse let me calculate in an entirely open-ended process, to be myself fully. I can see exactly as I please. Shapes are as in themselves they really are not. This is Wilde's critic/artist at work—as an artist/critic.

# EXHIBIT 3: PEDAGOGY

No one ever doubts the value of the studio when it comes to teaching art and design.[1] The content of the studio, however, isn't as secure. Usually, it's described in words, sometimes with visual prompts. This may be a step-by-step program or a well-wrought plan, or mumbo jumbo in a magical formula that sometimes helps to find yourself—each a segue into an open-ended (ambiguous) process with indeterminate (ambiguous) results.[2] And the studio is the same everywhere in between these extremes, from picky specification to powerful incantation—there are no right answers. In architecture, for example, the studio may rely on a version of the Vitruvian canon—a mix of firmness, commodity, and delight, adapted to the vagaries of the present. Ideally, these separate categories interact mutually, coadunate as they're defined and applied. But even in an impossible tangle, they may be ordered in value, one skewed from the others, or any ignored. Firmness and commodity put expertise and professional acumen first and foremost, and have a lot to say about problem solving and thinking things out—purpose, practical aims and goals, function and use. It's best to design for sustainability and social justice; many strive to make the world a better place. Delight, however, puts seeing and beauty first—nonetheless, what to see and how this looks and feels are ill defined and in large part, ineffable. Maybe that's why the studio seems emotionally charged, so that seeing is impressive whether you know what it is or not. Seeing defies explication and judgment, as long as sentiment in taste and beauty is personal, fickle and unrestrained. In studio teaching, there's no telling how to make (draw) a beautiful form, although it may be *de rigueur* for design. Whenever it's in play, delight is likely the overriding standard to measure progress in one-on-one desk crits and in pin-ups and small reviews, even if firmness and commodity or other things are key, and to assay results when the studio wraps up with everyone's designs on public display, no matter what the studio's stated aims and initial goals. Aesthetics (perception) is at the quick of art and design—is it any wonder that the eye always gets its way? Categories and

such—including meaningful nods and gestures, a pat on the back, sounds of encouragement and support—highlight success and sometimes less, but they miss the eye's
tricks in free flight to see as it likes. This is the root challenge in studio teaching—for
shape grammars, it's to show the eye at work in terms of rules in the embed-fuse cycle,
in tricks to reveal a beautiful form and to merge categories as they commingle in its
sweep. A surprisingly simple formula sets the goal in this kind of aesthetic education,
four words that deserve a single line—

Do what you see.

Visual calculating in shape grammars is the only way I know to add to the studio that
doesn't limit its sway—that's consistent with uncertainty (ambiguity) in what to see
and what to do, with no ironclad measure of progress or success. In "Seven Questions"
I show how the embed-fuse cycle unifies and re-divides (creates), to supersede lifeless
fancy in imagination.[3] This fills the soul in a burst of romantic enthusiasm—pulse after
pulse, heartbeat and breath, metabolic, organic, and vital.[4] Rules are recursive, but not
mechanical because of this; they do what I see with no letup, over and over again every
time they're tried. Seeing and doing reciprocate equally between embedding and fusing, the complementary faces of reduction (see Exhibit 1 for the math). The embed-fuse
cycle goes like so—

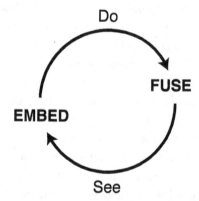

Of course, I'm not the only one looking—others in the studio can do what they see,
as well, individually and side by side. One is the same as two or more in concert or
competition, in small groups or large ones. The embed-fuse cycle ensures equal access
to shapes and their parts, in the kind of universal participation without restriction
that I talked about at the end of A1 in "Seven Questions." No one is privileged when

Exhibit 3                                                                 119

it comes to perception—what anyone can see, everyone can see, although some (artists and designers) may see more productively than others. There are no prerequisites to seeing—the studio is a democracy of the eye. Many years ago in the early 1980s, I used shape grammars for visual calculating with spatial relations, to recast the studio and its content in terms of Frederick Froebel's "kindergarten method" and its wonderful inventions for lively free play.[5] My method was simple—to combine the pieces (children's play blocks) in Froebel's building gifts, or in any vocabulary of shapes for that matter, in rules. The idea was to use rules in the embed-fuse cycle to augment Froebel's categories as the original source (generators) of new forms for the kindergarten, and to connect rules and categories in a kind of back-and-forth give and take.[6] There were forms of "life" (palaces, houses, tables, and chairs), "knowledge" (the Pythagorean theorem and its time-honored kith and kin), and "beauty" (symmetrical arrangements and patterns laid out on a flat surface, a table or the floor in the classroom, playroom, or kitchen). The categories worked for initial motivation and to suggest nice things to try in examples—drawings and models—but rules were something extra to add to the categories and to extend their reach, surpassing them in many ways. Rules put everything together in terms of the varied symmetries of the individual blocks in the building gifts, locally not globally in completed patterns; they made it practicable to order blocks by eye, in spatial relations that I could confirm by hand, using my thumb and forefinger to match corners and align edges in the proper way. I was never stuck—neither with too many ways to go on, making it impossible to choose, nor with too few and nothing much to try. All of this was sensory—seeing and touching. There was even a little children's rhyme, to include hearing in an audible track of blocks as they were combined in one arrangement after another; four lines to recite and then follow to make rules trouble-free and fun to use—

> Face to face put. That is right.
> Edges now are meeting quite.
> Edge to face now we will lay.
> Face to edge will end the play.

Rules produced wildly different designs, and these were easy to describe and classify in ways that Froebel would like, and in the mind's more exotic taxonomies, anything from building types to ironmongery to utensils and odd names for gadgets and gizmos (see above and in note 6). I thought this was pretty nifty—the endless variety and what I could do in the studio with spatial relations in rules was impressive—but my colleagues in art and design didn't think so, and they were adamant in their disdain. To be fair, some admired what they saw, only to dismiss my results once they discovered how they were obtained—the process was flawed, unfit for art, a profound embarrassment. Yes, my results were "permutations and combinations"—that's the way play blocks are when they're put together, objects that behave like points. Play blocks are counting out with units, fancy with original (given) counters in mechanical excess, that blunts imagination and drains esemplastic power. Today's computers are the same in AI and BIM. No matter, the rules I used were perfectly fine in the embed-fuse cycle—I made sure to emphasize this, although at the time, it went mostly unnoticed.[7] Then I was OK with permutations and combinations, as long as they didn't interfere with what I could see and do—in fact, I could use my method on polygons, in drawings of triangles and squares, etc. that were loaded with ambiguity.[8] Now I've lost my patience with permutations and combinations, and heuristic (computer) search in the immense design spaces in which their numbers are defined one after the other in mechanical certainty—what good is search when it's blind? (It's worth observing that justice, and law and order are cursed with a similar problem. Many agree that beauty is entirely in the eye of the beholder—and there's plenty to say about this and how it works in shape grammars and the embed-fuse cycle that's indifferent to contradiction as it pulses with new meaning, change after change. Nonetheless, real justice strives to be fair, even if this kind of consistency/uniformity dispenses with the inherent ambiguity in facts and circumstances, and the misdirection and misleading intentions of the law. When justice is blind, it gives up on a vital source of productive reasoning, and progressive methods that spur insight and imagination; then critical/legal reasoning, for example, in analogy, precedent, and rules, misses what the artist sees in rules of his/her own— the contradictions of the extravagant eye, it seems, exceed logic and law. See note 72 in "Seven Questions" for more on fairness and how it's achieved fully in AI.) There's nothing to see outside of the embed-fuse cycle that hasn't been seen before in the same way. Maybe this is a good time to be totally aboveboard about the embed-fuse cycle and how it's always been, in two correlative formulas—

What happens in the embed-fuse cycle isn't combinatorial.

There is no originalist (intentional) imperative in art, no mechanics (logic) of design.[9]

No one wants to demean the studio in the kindergarten—after all, the kindergarten is for children, and look at the things they do. Maybe that's why I'm still fond of my take on Froebel, and continue to think that it adds to art and design pedagogy, and to creative practice, as well. But rather than keep strictly to teaching that puts art and design in the kindergarten, stocked with play blocks like the ones that delighted me as a child, I'll extend teaching to schemas for rules, avoiding the prerequisite for a vocabulary

**Exhibit 3**     121

of shapes—that comes later, after the fact, as a consequence of calculating.[10] In truth, that's the only way I know how to explain the origin of vocabularies, unless God intervenes in advance with a vocabulary of his/her own, a gift to humankind, to solve this perennial mystery and so ensure that my eye is locked-in on divine constellations (arrangements of points). The gift of vocabulary, like each one of Froebel's building gifts in the kindergarten, makes visual experience safe and secure with few surprises; it keeps me from seeing whatever I like and changing my mind—every vocabulary is a wise teacher, and an inculcator of good habits (etiquette, manners, and tidiness) that impose a sinister kind of censorship in a strict standard to take away the looking. There's only God's canonical/conservative view in his/her perspective of eternity—too long for perception (seeing) in the here and now, and for rules in the embed-fuse cycle. I guess this spells the end of participation and a democracy of the eye. Does anyone put up with this loss in the studio, and art and design; is one way of seeing ever enough—even God's? I'd rather take the risk and see what happens in the embed-fuse cycle, without a given vocabulary; it's my choice, not explicitly all at once (only God can do that), but piecemeal in terms of what I see and do in an unfolding process—with schemas to help. There's no substitute for individual taste, calculating with rules. (Samuel Taylor Coleridge's two kinds of imagination in the *Biographia Literaria* may go for the uneasy relationship between God's vocabulary and the embed-fuse cycle—see note 49 in "Seven Questions." Even as God's "primary" imagination informs common perception, the artist's/critic's "secondary" imagination re-creates personal experience, to overcome/transcend God's everlasting influence and eternal dictates. What the artist/critic sees alters things as in themselves they really are—Oscar Wilde's aesthetic formula is a corollary of Coleridge's imagination.) The studio relies on a key idea—

Teach schemas before categories, including vocabularies.

And in a couple of analogues—

Schemas are for delight, and firmness and commodity; for beauty, and function and use, etc.

Seeing (perception) is magic—"The eye is the best of artists."[11]

Strictly speaking, schemas are categories—sets of rules that match given descriptions. Rules in schemas inform categories in art and design in many ways—from architecture's three to Froebel's matching ones (firmness is knowledge, commodity is life, and delight is beauty) to whatever makes sense now. Some of the ins and outs of this dot "Seven Questions." But the intuitive division I have in mind for schemas and categories does the trick for studio teaching. Roughly speaking, schemas stress process; they're productive suggestions to follow by eye. In contrast, categories measure results, in order to describe and classify them in qualitative/quantitative distinctions; categories encapsulate prior experience. At the very least, schemas are heuristics to probe categories and survey their reach. In shape grammars, there are no definite ends in categories, only variable means to go on in schemas. The ability to go on with rules in the embed-fuse cycle is the open sesame to the studio, and the secret to genius in art and design. William James captures the gist of this nicely, "Genius, in truth, means little more than the faculty of perceiving in an unhabitual way," that is to say, to see differently than you thought you would—to value what you see and to act on it. Why not teach this explicitly in the studio and in art and design—in how to do what you see now as the way to go on, instead of keeping to what you've already done? Why not surprise/surpass yourself? Who says you have to stick to your original intentions—to what you think you've been doing all along? Prior plans aren't binding and invariant; they can be changed or ignored at will. You're free to start over from wherever you are, whenever you please, without losing what you've already done—it's a great feeling to strike out on a new path, even if the fork is unknown, uncharted, and scary in a vast and inconstant region. Yogi Berra (another American philosopher) adds to the confusion—when you see a fork in the road, take it. But there's really no way to fail when there's nothing to stop you from going on. This follows the luxuriant eye as it swerves in its own special arc, and it's exactly what rules in schemas are for, as they apply recursively in the embed-fuse cycle. That's their use and value in the studio. I tried schemas for the first time in "Kindergarten Grammars" to define rules in conjugate sets (see note 5 in this exhibit). Deciding how shapes go together is a pretty good place to start; they're combined in the schema

$$x \rightarrow A + B$$

in which the variable x is either the shape A or the shape B in a spatial relation that's resolved in the sum A + B, with A and B both in a given vocabulary—Froebel was keen on the different pieces in the building gifts to combine in designs that belonged to his three categories. And then, there's the inverse of the schema, namely

$$A + B \rightarrow x$$

to take away A or B in the rules A + B $\rightarrow$ B and A + B $\rightarrow$ A. In fact, this is a universal format for rules in the embed-fuse cycle—for any rule A $\rightarrow$ B, with no vocabulary in particular in mind, I have A $\rightarrow$ A + B and the inverse A + B $\rightarrow$ B. (The compound term A + B simply cancels out.) Using these additive and subtractive devices is enough for a host of remarkable designs that unfold in terms of delight, and in parallel with description rules and weights, in terms of firmness and commodity (see above and in note 6)

**Exhibit 3**                                                                123

—delight is endlessly flexible for use and reuse, etc. But there are many of other schemas that suit the studio, as well—they're richer in aesthetic (perceptual) draw, heuristic
power, and intuitive sweep than the two I tailored expressly for Froebel in terms of spatial relations in pairs of play blocks A and B. There's more for the eye to see in schemas.
Let's do the rudiments instead of jumping headlong into trickier stuff—the eye is sure
and knows the way. A schema

$x \rightarrow y$

is a set of rules

$g(x) \rightarrow g(y)$

that are defined when shapes are given as values to the variables x and y in an "assignment" g. For example, when g(x) is the shape

and g(y) is the shape

I get the familiar rule

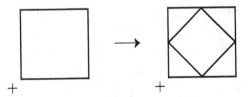

with no telling how these shapes are divided in the rule—they're only drawings with
no fixed analysis in words or otherwise. Maybe g(x) and g(y) are given parametrically,

or taken from other shape grammars with their own schemas and rules—either way, this also leaves the shapes in rules unanalyzed. A nice way to define schemas is to express the variable y in terms of the variable x. It's the method I used for the three primary schemas in "Seven Questions." By name, there are schemas for parts, transformations, and boundaries—

(1) y is part of x

   x → prt(x)

(2) y is a transformation of x

   x → t(x)

(3) y is the boundary of x

   x → b(x)

If g(x) is the triangle

then the primary schemas include the three rules

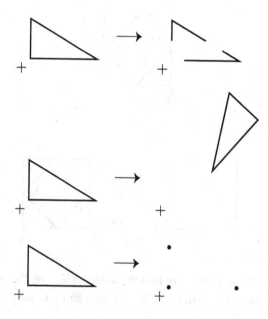

Exhibit 3                                                                    125

where the part of the triangle in the first rule varies from the empty shape (a blank space) to the triangle itself, and t in the second rule is any transformation you choose, from isometries and similarities to projections and nonlinear variations. I can use the primary schemas to get all sorts of others—useful heuristics for this include subsets, copies, inverses, adding, composition, and Boolean expressions. Of more generality, any two algebraic expressions in symbols, with any number of variables, are good for the left-hand and the right-hand sides of schemas—these expressions can be defined mutually and not, as long as shapes are produced in assignments. I can also define schemas in computers in AI/BIM with data and parametric representations, etc.—in fact, parametric representations are something I suggested before. Individual shapes related in rules are readily given in descriptions like John McCarthy's for a vertical line connected to a horizontal segment in A3 of "Seven Questions," and like the more elaborate description for A's with hinged legs near the end of Exhibit 2. For architecture, and design construed broadly, I can define schemas for different grids as the generators of forms—to measure rhythmic patterns and to place shapes with respect to one another. (None of this means that AI/BIM is suddenly good for seeing, and art and design—it's not. Classifying rules in schemas isn't the same as using rules to calculate in the embed-fuse cycle. As categories, schemas are forever—what rules do is always evanescent.) Or less formally, I can stick to words and pictorial examples, when everyone agrees that this is OK. Paul Klee is pretty good with words and drawings throughout his *Pedagogical Sketchbook*, especially in the opening exercises for lines that combine in terms of lines—both visible and not.[12] This makes it easy to show what to do without too much detail. To see how this goes in an analogous way in rules, let's try it like so— in one quadrilateral (the left-hand side of a rule) inscribe a second quadrilateral (the right-hand side of the rule), vertices to edges, in the schema

   quadrilateral1 → quadrilateral1 + quadrilateral2

for rules that look like this one

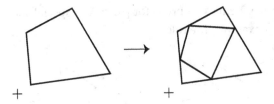

in a generalization of my rule

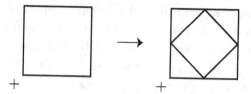

that inscribes squares in squares, but that answers to many other descriptions, as well. I can turn my rule for quadrilaterals into a schema by associating variables for the coordinates of endpoints with given conditions, as expected in analytic geometry—in a full-fledged parametric definition of spatial relations for indefinitely many pairs of quadrilaterals. (Doesn't this show how the schema x → A + B works for spatial relations in the embed-fuse cycle?) Other kinds of dimensioning given in terms of variables are good, too, maybe from engineering design, although sometimes drawings with coordinates and such can be really ugly, obscuring shapes in the rush of detail. And notice that the schema for quadrilaterals includes rules like this

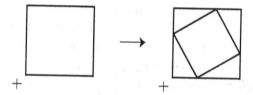

in which squares are related in different ways. The aim is to keep schemas simple and intuitive; defining them isn't a logical game of proof, to derive expressions rigorously from a handful of clever axioms—any two shapes make a rule, no matter how you get them, in drawings, symbols, words, etc. As I confessed in Exhibit 2, note 9, my all-time favorite schema is for the identities

x → x

that's a subset of the schema x → prt(x) for parts. And special subsets of the identities are often useful; for example, the rule

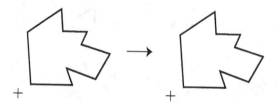

Exhibit 3                                                                127

is in the identity schema for polygons—defined formally or in words—along with identities for triangles and squares, etc. The schema for parts defines all rules, too, either in
the composition

$x \rightarrow \text{prt}(\text{prt}^{-1}(x))$

that incorporates the inverse of $x \rightarrow \text{prt}(x)$ in a neat variation of the kindergarten schemas $x \rightarrow A + B$ and $A + B \rightarrow x$, or commutatively in the composition

$x \rightarrow \text{prt}^{-1}(\text{prt}(x))$

All rules are also in the summation schema

$x \rightarrow \Sigma \, F(\text{prt}(x))$

But this trio of universal schemas is already familiar from "Seven Questions" and its
notes. What I want to show now is the visual sweep of schemas, when different values are assigned to their variables to produce different things in the same way. This
establishes the worth of schemas in the studio to encourage a truly vast—in fact, an
inexhaustible—array of designs from teachable (common) means that track what the
eye sees, as it swerves from this to that. For this purpose, I'll start with the identity
schema

$x \rightarrow x$

Next, I'll try the addition schema

$x \rightarrow x + t(x)$

that keeps the formal machinery of $x \rightarrow \Sigma \, F(\text{prt}(x))$ to a bare minimum.[13] Easier still,
this merely adds the identity schema $x \rightarrow x$ and the transformation schema $x \rightarrow t(x)$,
in the same way my schema $x \rightarrow A + B$ in "Kindergarten Grammars" adds two schemas
$x \rightarrow A$ and $x \rightarrow B$ with constant right-hand sides. And my rules that inscribe squares
in squares, and quadrilaterals in quadrilaterals also do the trick. Then to finish up,
I'll augment shapes to explore color variations in many shades of grey, in terms of
weights, and the inverse of the boundary schema $x \rightarrow b(x)$ in the coloring book
schema

$x \rightarrow x + b^{-1}(x)$

that's a subset of the synoptic sum and composition

$x \rightarrow x + b^{-1}(\text{prt}(x))$

(The schema $x \rightarrow x + b^{-1}(\text{prt}(x))$ is a self-indulgent flourish, ornamental symbols to
show what's possible from a simple start—once you get going with schemas, this is

awfully hard to resist, and I'm not one to avoid wanton temptation in art and design. I said so in "Seven Questions," with many examples.)

The varied identities in the schema x → x leave shapes alone as they alter how they look. The trick is to make new designs entirely by eye, seemingly aloof and physically disengaged, not getting your hands dirty—design without end, seeing in an aristocratic kind of magic in which shapes change without ever changing them. (Wilde's critic is keen on identities when he/she is an artist—and every great critic is.) For example, the shape

looks different as two squares for the identity

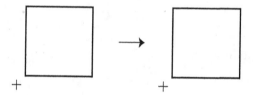

than it does as four triangles for the identity

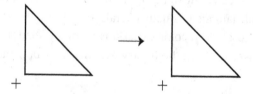

In fact, dthat[the philosopher I see standing to my left] in note 73 of "Seven Questions" couldn't stop talking about how neat it felt to change what he saw at will, looking at the same shape, calculating all along—identities are tricks worth knowing in art and design, ones that open untold possibilities. The late Jonathan Miller was there, too,

**Exhibit 3**                                                    129

attesting to such tricks in opera, and elsewhere in the arts. In the same vein, identities show why copying is a creative activity—I can copy a given parti

in starkly contrasting follies

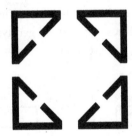 

Architecture teems with instances of this kind of unapologetic plagiarism that bursts with esemplastic power and creativity—fusing and embedding. The ability of identities to re-divide and parse shapes in a dynamic process, to change them and charge them with meaning in order to alter visual experience in myriad ways, is the essence of imagination—in fact, of imagination and conscious understanding/parsing in free play. This is evident in a host of varied examples, in whatever catches my eye long enough to pause and look. Take the shape

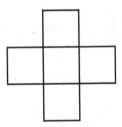

I can try identities for polygons in these assignments

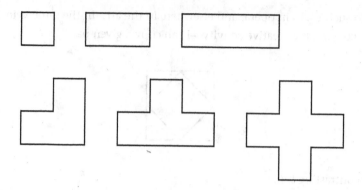

to see it in a bunch of alternative ways. If I keep to discrete polygons (figures without any parts in common—endpoints aren't parts of lines, etc.), then I get five decompositions that I can represent conveniently in graphs, in particular, in trees—

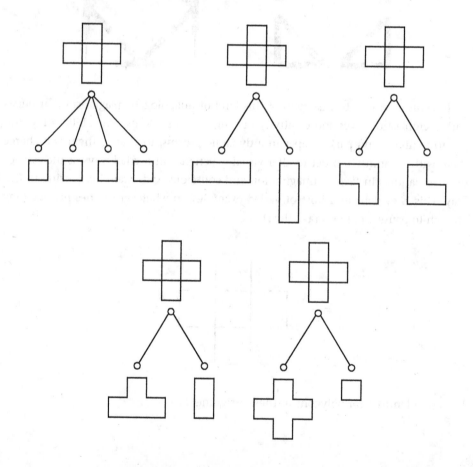

**Exhibit 3**                                                     131

If I use geometrically similar (congruent) polygons instead, I get five new results in these eight trees

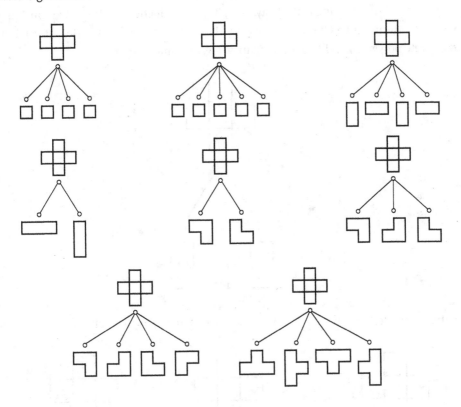

And, of course, I'm free to ignore these provisos completely, to get a Greek cross, a short rectangle, and a long one

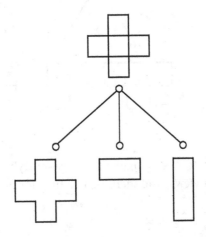

For both discrete polygons and similar ones, some of the trees are unique and others are instances of trees in which polygons are rotated and reflected in a dihedral symmetry group. That's why counting isn't always 1, 2, 3, . . . . (I touched on this in Exhibit 1 and Exhibit 2 in terms of La Grange's theorem.) There are two trees with different versions for discrete polygons, and two for similarity. For example, the tree

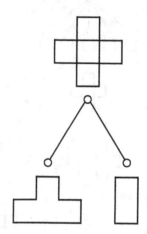

for the T, inverted in this case, and the discrete rectangle has three alternatives

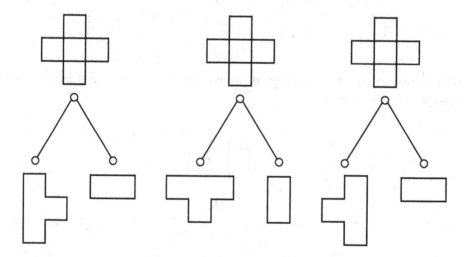

allowing for four layouts to be distinguished in use—maybe one leads to a building plan with better access or views. And there are two trees for similar L's that are discrete, as well—

**Exhibit 3** 133

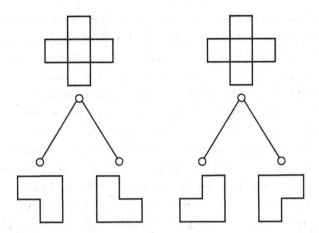

How many versions are possible for the tree

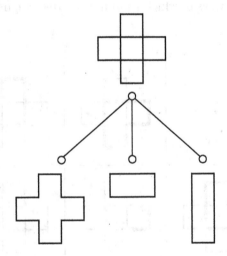

for the Greek cross and a pair of different rectangles? Try it for the Greek cross with a square in the middle, and pairs of squares in its arms—for trees with four smallest polygons. I guess it's the same for discrete pentagons on the diagonals or the horizontal and vertical of the shape

Identities allow for shapes to be compared in different similarities and other relationships—the Greek cross is a funny pentagon in a visual metaphor. (This may be
egregious with respect to the Greek cross and the shapes that alter it, and may lead to
heresy and even apostasy if dogma claims the cross and limits how it's used. Identities in the embed-fuse cycle are indifferent to this, in accord with Coleridge's thesis,
antithesis, and synthesis. Shapes are vital, like living things—they assimilate whatever artists and designers, and anyone else happen to say about them. Of course
in actual practice, there may be dogma and heresy in the studio; the studio isn't
beyond orthodoxy, and it's influence may be good and bad. Nonetheless, the studio goes on just the same, independent of belief and faith—I can see what I please
in spite of dogma, no matter how firmly it's established in liturgy, ritual, etc., and
do what I see. That's the covenant identities make, and every other schema, as well.
In fact, the very possibility of heresy and apostasy proves it.) Sometimes it pays to
reconcile contrasting trees in synoptic graphs that resolve a common vocabulary of
shapes and parts. The ways in which L's in my two trees align/interact are shown in
this table

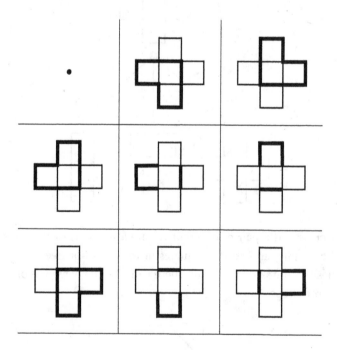

to define the graph

**Exhibit 3**                                                                           135

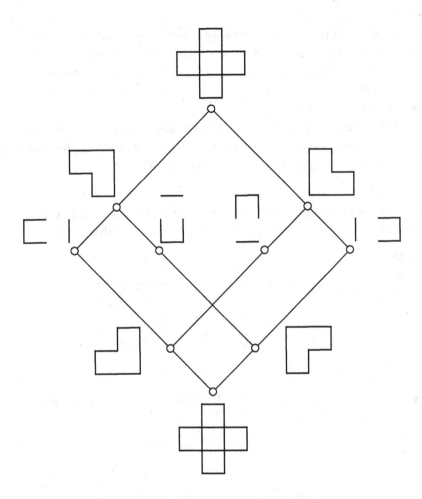

in which L's are divided, so that two in one way make the two in the other, shuffling parts around. A nice way to think about this is that there are shapes in the table, and special symbols (names) in the graph—going from concrete experience in shapes to abstract structure in symbols. Seeing in multiple ways seems effortless—it's merely looking—but this takes some work to explain, at least in terms of common parts. Many find this very satisfying on the path to personal understanding—graphs show something important about underlying structure and how it unfolds. Try to merge some of the other trees I've shown to see what vocabularies are defined—keep it simple, graphs can get complicated fast to put them outside the easy reach of the eye. But really, defining graphs is an idle exercise—the embed-fuse cycle makes descriptions and representations of shapes in terms of fixed parts (counters) entirely unnecessary. There's no vocabulary of shapes that works everywhere or every time—that's the reason for visual calculating in shape grammars, and why it's vital, for art and design to stay with the eye as it swerves in its own way. It's easy to agree that identities are creative for the artist and the critic alike—in fact, for the teacher in the studio, as well—in what they all do as a simple matter of course.

But what about adding to shapes, to get new designs in a more conventional way, where copying and plagiarism aren't an immediate/likely barrier? This also relies on seeing in the same way identities do, in the addition schema

$$x \rightarrow x + t(x)$$

for different transformations t and assignments to the variable x. In fact, the identities are in $x \rightarrow x + t(x)$, when t is an identity transformation, so that $x + t(x) = x + x = x$. Moreover, the rule

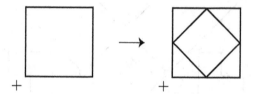

is in $x \rightarrow x + t(x)$, for a square, and rotation ($\pi/4$) and scale (1/2). Defining such rules is no big deal. I can try the rule

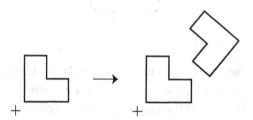

for a pair of dancing L's like the ones in the preface, in an aleatory whirl—some like to design in this random sort of way—along with the conjugate rule

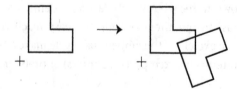

**Exhibit 3**                    137

If I start with a single L, I get

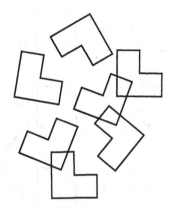

among many alternatives. And for squares, there's this

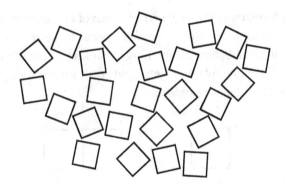

from this

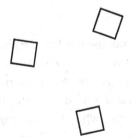

for the trio of rules

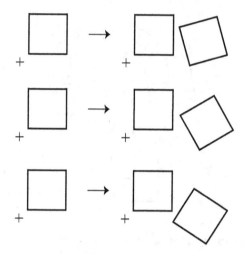

in which the transformations are small perturbations of one another, that make squares jiggle and wiggle around. But things needn't appear to be so unruly; the addition schema also gives me well-behaved, tidy things, that keep to rules in the way everyone has come to expect, for example, in regular grids with squares and triangles that fill a sheet of paper. Each grid takes a single rule

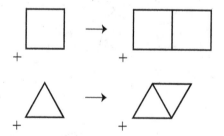

in a boustrophedon trace—but no one wants to be an ox. Nonetheless, the same goes for geometric tiling of all sorts—M. C. Escher has some famous examples. (Using shape grammars to define grids and tilings provides an alternative to computers and AI/BIM. Often enough, shape grammars and computers match up, with no improvement to the latter—but this was the point earlier, when I was talking about the use of computers for parametric representations of grids and such.) There are many things to do that make a

**Exhibit 3**                                                                                      139

difference. I haven't said very much about the transformation t in the addition schema x → x + t(x). It's another good way to bring in variety—maybe t alters according to given parametric expressions, as a value or values are assigned to x. (And again, computers are a must for this in parametric design and BIM, only to take away the looking in the embed-fuse cycle.)

Finally, at least for the time being, I can try filling in, in terms of the coloring book schema in the simple form

$$x \rightarrow x + b^{-1}(x)$$

defined for weights, shades of grey that get lighter and lighter whenever they overlap—who would ever think of such an odd inversion? Weights are coequal properties of shapes, with their own parts to add and take away, and transform. Shapes are arrayed (augmented) freely with anything from labels and symbols to line thickness and texture to colors and materials. For example, labels may belong to sets with subsets and unions, etc., or concatenate in strings (series) that answer to formal/mathematical expressions—labels make it a cinch to count with tally marks, and recursively, to add and multiply, etc. But let's not go too far astray, let's go with grey—take the square

and use the coloring book schema to fill it in

If I do this twice, three times or more, I can change grey tones as in overpainting. And for two squares, one inside the other

it looks like this

for the three different parts that form boundaries, and like this

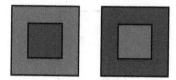

in combinations that reverse grey tones, when I try my rules more than once. Of course, I can go on with the shape

in the same way, for its parts that form boundaries

**Exhibit 3**                                                                                  141

and in varied combinations

Possibilities multiply exponentially by 2—try it for four squares in the shape

and its parts. (I know I said this would be the end of it, but I can't resist going on to reprise a more elaborate example of coloring from the early days of shape grammars, sometime in 1968 when I was doing a lot of painting—the whole nine yards with brushes and pigments, and big canvases. Simplifying something that's too complicated to start is a nice way of showing the path you're on. It's neat how some things seem to get easier as time goes on. Try the inverse of the boundary schema

$$x \rightarrow b^{-1}(x)$$

for F in the summation schema $x \rightarrow \Sigma F(prt(x))$ to get the picture

I put in, in "Seven Questions"—start from the underlying cartoon

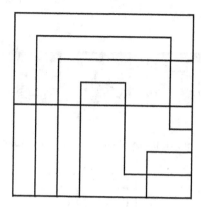

This line drawing relies on the addition schema $x \rightarrow x + t(x)$, but combining greys in areas—overpainting in terms of weights to relate distinct regions—is a generalization of my original use of Venn diagrams and painting rules in "material specifications."[14] There's a brute force way of doing this for thirteen discrete areas, twelve of them multiple times, and an easy way with four overlapping areas. Moreover, there are other ways in between, for example, à la Venn in material specifications. The easy way is worth a look in sequence—these four areas

combine successively like so—

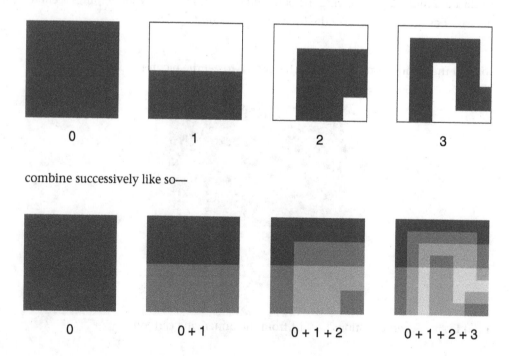

Exhibit 3                                                                143

in an extension of silk screen. And other sequences work, as well—including mine, there are a couple dozen in a kind of confluence. Try a few, at least the one that goes in reverse. This isn't problem solving in which computer methods like divide and conquer apply in recursive subdivision, but open-ended seeing that proves the magic of the artist's eye in the endless tricks that play out in $x \rightarrow \Sigma\ F(prt(x))$. Neat, but I don't want to imply that everything should be expressed in parallel, in a single rule application—this may elude the eye, with too much to see that comes all at once. Sometimes, it's better to unpack things separately. It may reveal more when they're seen leisurely in a longer sequence, as they alter in smaller, more manageable steps, part by part over time—and in fact, my picture is an example of both parallel and sequential improvisation. There's no best way to see; this is always up for grabs.)

Schemas allow for all sorts of cool stuff—I can see in many ways, copy without theft, add to what I've got, complete and take away, fill in at will, make any distinctions I please with labels and colors, change my mind without rhyme or reason, and go on and on. This tracks pretty close to the categories in Harold Bloom's six revisionary ratios for poems, and Nelson Goodman's speculative ways of world making (see note 65 in "Seven Questions"), only to veer from both in the formal details in Exhibit 1 that make schemas work in math—schemas let me design, and they let me extend this to teaching in the studio, as design unfolds fully in beauty and delight, and in concert with firmness and commodity. Schemas are part and parcel of the creative process in the embed-fuse cycle, in visual calculating past fancy, with rules for imagination and esemplastic power. Shape grammars are the reason why; otherwise, calculating is no more than counting, Sisyphean coding and debugging that give up on seeing in a quest for certainty—0's and 1's in "fixities and definites," always 1, 2, 3. Seeing and doing in rules are vital—the source of beautiful form in art and design, and the origin of delight. In schemas, there are myriad myriads for the eye to see, so that there's always plenty more to do.

This exhibit is far from being exhaustive—it's meant merely to suggest the range of possibilities in a few handy schemas. It will have served its purpose if it gets you to think schemas through on your own, to try them out in the studio or even in the classroom on the blackboard, or to use them in practice as a way to design—this can be your secret method that no one else knows. Schemas and how they're defined aren't hard to keep in your head, to try here and now and again, as heuristics and prompts for when you're stuck and don't have a clue where you might go next. Schemas show design in action—as the eye swerves, I do what I see. And there are real world examples of this in the studio. Donald Schön traces one in architecture in *The Reflective Practitioner*—nearly 40 years after its publication, it's still a focus for lively discussion in my research seminars, along with Herbert Simon's earlier *The Sciences of the Artificial*. It seems that progress is glacially slow trying to make sense of art and design—schemas in the embed-fuse cycle aim to fix that, when they put calculating and imagination together to interact mutually. No doubt, imagination is the missing ingredient—Coleridge thought so 200 years ago, and he's still waiting for his recipe to catch on. Taste is hard to understand. Schön traces the back and forth in Quist's studio—Petra goes from a flawed ground plan

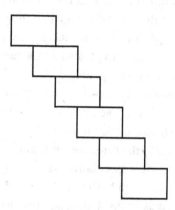

in which rectangular areas are too small, to this one

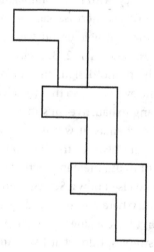

with one-third more interior space.[15] Rules in schemas unpack what Petra does as they intercalate the missing details in a visual process that goes step by step, calculating

**Exhibit 3**

145

in the embed-fuse cycle. Petra may well deny this, but she probably doesn't know what she's up to—isn't that what Quist is trying to show her, at their regular desk crits without the benefit of schemas? Let's see what schemas add. The transformation that's given in the addition rule

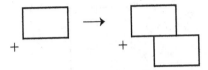

in $x \rightarrow x + t(x)$, to define the plan

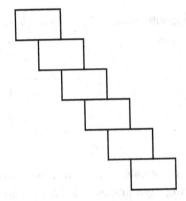

in terms of Petra's rectangle is changed in the two addition rules

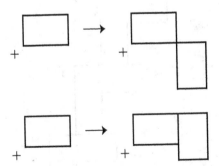

for the alternative layout

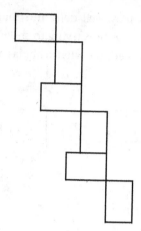

among a host of varied possibilities—see what else you can get. But keeping to what I
already have, the transformation

in the primary schema x → t(x) applies to turn three wasteful outside niches into three
useful inside corners, flipping right angles to form the trio of staggered L's in Petra's
improved ground plan

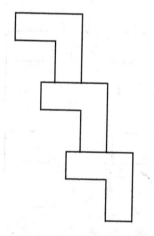

**Exhibit 3** 147

Presto chango—simple enough when you do it by eye. And if Petra complains that this takes too long—that she did it in an instant, all at once in her head—she can try it in parallel, using the summation schema x → Σ F(prt(x)) for the rectangles in her original ground plan.[16] It's straightforward—use the transformation schema x → t(x) to move each of the six rectangles to get the horizontal and vertical ones in my intercalated plan, and open up the horizontal and vertical rectangles like this

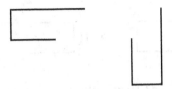

to go all the way to Petra's solution. For example, the rule that's used for the topmost rectangle in x → Σ F(prt(x)) is this

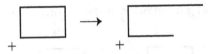

and the rule for the next one down is

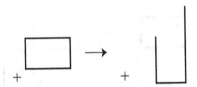

The rules for the four remaining rectangles are straightforward, too—align the six rectangles and the three L's, so that the top rectangle and the horizontal arm of the top L coincide, and then draw (trace) what you see. (These rules are in a comprehensive schema x → t(prt$_1$(x) + t'(prt$_2$(x))), where x is Petra's rectangle, prt$_1$(x) and prt$_2$(x) are discrete and sum to x, and prt$_1$(x) + t'(prt$_2$(x)) is the shape

OK, even if the details are way too fussy—they need a blackboard and chalk to explain. Otherwise, the schema is easy to mistake for a strong password on a computer. (The risk of symbol porn is discussed in note 12.) Luckily, there are easier ways—maybe symbols and words are enough in the schema

Petra's rectangle → t(Petra's open rectangle)

And why not replace the words with a couple of drawings

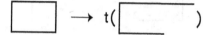

that don't need to be described? Then, only four symbols are left in the schema— →, (, ), and t.) Not too shabby, although pretty farfetched. Petra's boast is bigger than she thinks—but the mental gymnastics will change her mind. Seeing things unfold in terms of schemas—to do what you see on the fly, instead—may be a better way to get the results you're looking for than thinking things out in full before you act. There's a lot more for Petra to try in ground plans, with the three addition rules

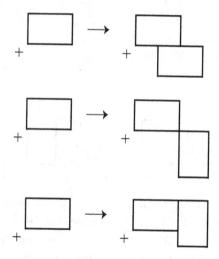

in the schema x → x + t(x)—maybe Petra can associate points with rectangles to better understand this process in terms of their fourfold symmetry (see note 5 in this exhibit for the details of the method). Or maybe Petra's improved plan strikes like lightning out of the blue, in a gratuitous sort of way—and there's a rule for this, too, although there's

Exhibit 3                                    149

not much in it to see, simply one complete plan going to another complete plan. Does Petra have a catalogue of plans hidden away somewhere that she consults when she needs it, a kind of Delphic oracle for design? How does she know what to choose and how to copy what she sees? Aren't plans, Delphic and not, notoriously ambiguous—it's probably safe to assume that Petra has the identities x → x hidden away, too, but is she ready to use them for help? Come on, what really happens when Petra has to do this another time, when she's finds herself in a blind alley and hasn't a clue how to go on? Poor Petra—isn't that the reason why she's in Quist's studio, to learn how to go on in a terrible jam? During their desk crits, Quist suggests a number of different ways out, and sometimes one them does the trick—maybe this helps Petra change little rectangles into bigger L's, and maybe it doesn't. Either way, it's where schemas come in, to avoid jams with the artist's formula—

Follow your eye.

It's also telling to see how Petra's plans work in the kindergarten, with my intercalated layout. The three addition rules for plans, one for Petra's original layout and two for the intercalated one, are in the schema x → A + B, where the rectangles A and B are congruent in different spatial relations

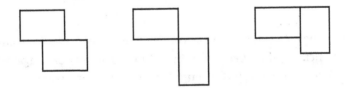

My test by hand using my thumb and forefinger, however, only allows for the last spatial relation in the series—bad luck for me. I guess I was too strict in my kindergarten, even if this can be fixed—new spatial relations can be defined in terms of ones like mine, when they're used to align and measure rectangles and other shapes, adding and taking away.[17] But this process seems obscure to most, requiring a few too many steps and lacking heuristic appeal—it seems a little abstract, although it's entirely visual, and sometimes it doesn't work. Quist's and Petra's drawing tools tell a more familiar story—the draftsperson's T-square and compass, triangles, and rulers and scales extend farther than the hand's silken touch, to what the skilled architect knows automatically in practice. These tools are the source of new ideas—in fact, a host of new spatial relations, including the three above. As long as it doesn't get in the way of seeing, technology can be a wonderful thing. And then there's the transformation rule

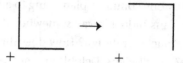

that turns niches into corners to make the three L's in Petra's final plan. This takes more than the rules in x → A + B. The schema is good for the rule

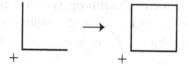

that adds a niche and a corner, but I also need its inverse A + B → x, for the rule

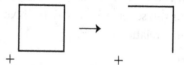

that takes away the niche and leaves the corner—the schemas x → A + B and A + B → x are linked together. Another spatial relation defines the rules—once again A and B are equal shapes, but this time each is two right angled sides of a square

In this spatial relation, A and B are members of a set—they behave like points rather than line drawings in order to distinguish the niche A and the corner B. Of course, the embed-fuse cycle is required for any of this to work, without an expanded vocabulary that includes divisions of Petra's rectangle to segment its sides, in the same way I tried to divide squares and triangles in the third example in Exhibit 2. And there's more to it, as well, if I continue to add niches and corners for decoration and ornament, or for function and use—maybe to individual L's on their diagonals

**Exhibit 3** 151

so that area is conserved, or across their diagonals in this way

Try it on the arms of the L's, on their concave sides or their convex ones. Or inflect walls

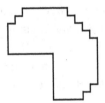

There's no stopping, when there's something new to see and do—rules in the embed-fuse cycle make this a sure thing. That's it for permutations and combinations in the kindergarten; they fade away, as soon as seeing overtakes counting, and imagination supersedes fancy—vocabulary aside in Froebel's building gifts and kindergarten play blocks, etc., the schema x → A + B and its inverse are a perfectly general way to calculate in the embed-fuse cycle. With no vocabulary to block the way, you're free to do what you see. I guess it's fitting to finish off with a drawing of Petra's unassuming rectangle

It's her gift to Quist's studio, not as vocabulary but as an open invitation to the eye—to soar joyfully in free flight.

Schemas do a lot, even from a modest start. It strikes me that the use of schemas can't but add to the impressive power of studio teaching. Schemas are something of lasting value to take away from the studio, that keep to art and design for seeing and doing in the embed-fuse cycle. Aren't schemas what the studio is really about—on the one hand, schemas are implicit (tacit) in line with categories in the Vitruvian canon, and on the other hand, they're explicit in terms of rules for visual calculating in shape grammars. And in both cases, firmness, commodity, and delight interact mutually, in any of their many versions, coadunate in a comprehensive view of design. I've been careful to emphasize that there are reciprocal relationships for schemas and categories. Nonetheless, schemas can be used for art exclusively, simply to do what you see without worrying about use or what anything is for. Everyone is quick to agree that the studio is an alternative form of education that's tried and true—there's no replacing it. And it makes good sense to try and extend the studio to include STEM in addition to art and architecture, etc., but this isn't for the usual reasons. It's not because art is instrumental in STEM subjects—looking at *Las Meninas* like Pablo Picasso or listening to Mozart in the background doesn't make you any better at math or a skilled engineer. Nor is it because art generates STEAM. Going from STEM to STEAM isn't a phase change—adding art for goodwill and therapy, R & R and mental health doesn't improve your life or make you a better person. None of this renders STEM more effective than it already is. None of it rescues art, rehabilitates it, or guarantees its inevitable role in STEM education. None of it matters. Yes, art can be related to useful things, but this isn't necessary—

Art doesn't need to be useful; it's vital in its own right.

Art and design in the studio and teaching prove the tricks of imagination—in terms of schemas, and rules in the embed-fuse cycle. Where else can you see imagination in action—again and again every time a rule is tried? Art (seeing) subsumes STEM (counting) in bursts of esemplastic power and creativity, beyond fancy in rules outside of logic and statistics, axioms, prior data and facts, and rote technique. Art follows the eye without the combinatory paraphernalia of fancy, and computers for AI/BIM that block the way in descriptions and representations—this is where the studio ends. (No doubt, AI/BIM will continue to hold vast sway, at least for the time being. In some precincts of art and design, new technology is always the rage, for example, in architecture, where AI/BIM promises algorithms and machine learning to drive design, in a cache of prior data that describes the way we really are, and what we need and desire. But like most fads in architectural education, this is likely to fade away quickly—the usual half-life is two or three years, maybe a little longer. The trick to design isn't in a technology that

**Exhibit 3**                                                                 153

takes away the looking. Computers and AI/BIM are blind; they aren't the key to design. Without seeing, architecture isn't art; architectural education is absent its strongest trait, the one that marks it uniquely. Then, it's only technical education with expert/professional answers in a brave new world—aesthetic education looks for more.) The embed-fuse cycle in shape grammars isn't calculating in the usual way—the usual way in Turing machines and computers is a special case. There are no prerequisites for imagination either in logic or statistics. Neither axioms nor data and facts are a substitute for the eye. The key idea for the studio still goes—

Schemas are for doing what you see.

# Notes

## SHAPE GRAMMARS: SEVEN QUESTIONS AND THEIR SHORT ANSWERS

1. A. M. Turing, "On Computable Numbers, with an Application to the Entscheidungsproblem," *Proceedings of the London Mathematical Society* (1937): 230–265. Some of my citations may seem redundant to computer experts, or to artists, critics, and designers, although rarely to both groups at the same time. Whenever I cite classics, it's usually in an odd way. My goal is to set the stage for shape grammars in order for calculating to include art and design, not to be trendy or fashionable. There's already plenty of that elsewhere. In theoretical computer science today, Christos Papadimitriou among others traces a similar arc; in sync with Turing, he focuses an "algorithmic lens" on the sciences and like subjects to show that key ideas as in themselves are calculating—special algorithms in economics, biology and evolution, recent linguistics à la Noam Chomsky, etc. But this computer perspective in which calculating provides a method of discovery and elucidation goes mostly if not entirely one-way—what about subjects that aren't close to the sciences, where key ideas as in themselves are not calculating, or seem not to be? Are calculating and computers at root incomplete, so that this division (digital gap) is permanent and a binary measure of value that separates two cultures? I'm always surprised how quickly calculating assays science and what's not. Nonetheless, the ordering is up for grabs; it's your choice to make—after all, science may be fool's gold. Or do subjects outside of calculating augment it somehow, so that it assimilates their additions and grows, and thereby offers a better lens to focus on more? This kind of mutual relationship is what I'm trying for, for visual calculating in shape grammars, and art and design—once art and design extend calculating past counting and symbols (descriptions and representations in 0's and 1's) to seeing and shapes, calculating adds to art and design. Calculating is complete, and without nuance, no assay for science and what's not.

2. The quotation is from Mary Shelley's "Introduction" to the revised (reanimated) 1831 edition—M. Shelley, *Frankenstein: or, The Modern Prometheus*, ed. M. K. Joseph (Oxford: Oxford University Press, 2008), 9.

3. In effect, the present prefigures the past. The rules I try alter the past to sync with the here and now. In retrospect, everything is given in advance; what I see is there before I see it, but afterwards—yesterday is decided today and the day before yesterday, yesterday, etc. How the present frames the past to suit itself is highlighted in G. Stiny, "Shape Rules: Closure, Continuity, and

Emergence," *Environment and Planning B: Planning and Design* 21 (1994): s49–s78, and the use of identities x → x for this is in G. Stiny, "Useless Rules," *Environment and Planning B: Planning and Design* 23 (1996): 235–237. I'm keen on the idea that doing nothing in identities is constructive throughout—remarkable tricks abound in seeing's idle ways of making.

4. J. von Neumann, "Theory and Organization of Complicated Automata," in *Theory of Self-reproducing Automata*, ed. A. W. Burks (Urbana, IL: University of Illinois Press, 1966), 46–47. There's little doubt that von Neumann puts the Rorschach test outside of the scope of visual analogies. (For additional proof, see note 21.) From my point of view, every shape is a Rorschach test, and can be described (interpreted/represented) just as freely—for any given shape, there's no single visual analogy (spatial relation) that goes for all the rest of it. A triangle isn't always three edges/sides. In this sense, every shape is inexhaustible; there's something new and different in it for everyone now, and many times over. Seeing supersedes description for all of us, throughout art and design.

5. G.-C. Rota, *Indiscrete Thoughts*, ed. F. Palombi (Boston: Birkhäuser, 1997), 58–59. Rota's Socratic dialogue with Stanislaw Ulam isn't to be missed. Somewhere in the middle, Rota poses the crucial question, that usually comes with opposing answers—one goes for computers (counting) and the other for seeing (phenomenology) in art and design. The dialogue runs so—

> "Do you then propose that we give up mathematical logic?" said I, in fake amazement.
> "Quite the opposite. Logic formalizes only very few of the processes by which we think. The time has come to enrich formal logic by adding some other fundamental notions to it. What is it that you see when you see? You see an object *as* a key, you see a man in a car *as* a passenger, you see some sheets of paper *as* a book. It is the word 'as' that must be mathematically formalized, on a par with the connectives 'and,' 'or,' 'implies,' and 'not,' that have already been accepted into a formal logic. Until you do that, you will not get very far with your A.I. [artificial intelligence] problem." . . .
> At this point, Stan Ulam started walking away, and it struck me that he was doing precisely what Descartes and Kant and Charles Saunders Peirce and Husserl and Wittgenstein had done before him at a similar juncture. He was changing the subject.

Of course, seeing an object *as* a key is to pick out the object at the same time. Or is it seeing *as* when you pick out the object, and then see it *as* a key? In shape grammars, it's a wash—both are OK, or is this retrospective, as it is for the topologies, etc., a little farther on in my answer? And changing the subject is also seeing *as* in another way—maybe there's more to embedding in shape grammars and visual calculating than anyone knows, where it's changing the subject, seeing *as*, every time you try a rule. No matter, walking away with René Descartes and other greats of philosophy makes for some pretty good company, and they'll figure it out—together with Richard Rorty's "ironist," for whom changing the subject is a literary skill, going fluently from one terminology to another in surprising gestalt switches. (I refer to Marjorie Garber's gloss on metaphor in terms of figure-ground reversal later in A4. There's more on Rorty's ironist and his/her use of "massive redescription" alongside inference in *Shape*, 153—see note 69 for the complete citation.) New perception is changing the subject as artists and designers readily do, whenever they chance to look again. Seeing is seeing *as*, as rules apply in an ongoing process. This goes for visual calculating, and for art and design, but for logic and the computer—well, maybe not quite yet. In fact, this has probably never been a goal—neither logic nor the computer offers very much encouragement to the ironist, with his/her negative capability. I guess seeing *as* in this way or that is OK a few times, to store the results in computer memory for subsequent use, but no more than five or six, and better two or three. This makes visual analogies possible, by leaving out a whole lot. What happens to the Rorschach test?

6. This section refers to bits and pieces of Owen Barfield's discussion of participation and figuration in *Saving the Appearances: A Study in Idolatry*, 2nd ed. (Middletown, CT: Wesleyan University Press, 1988). The two dot "Seven Questions," along with his distinction that divides original participation and final participation (collective representations). The poet and the critic alike admire Barfield's work, somehow including T. S. Eliot and Harold Bloom. In my preface, I distinguish shapes and symbols via their dimension i—i $\geq$ 0 for shapes, and i = 0 for symbols. The same taxonomy goes for Barfield—i $\geq$ 0 for original participation and i = 0 for final participation, in the straightforward analogy

original participation : final participation :: shapes : symbols

This ties in the analogies for shapes and symbols I've already given, and suggests new ones—maybe for the ratio of appearance to structure that also aligns with Barfield. For many, structure is the measure of cognitive scope, but without shape grammars, it spells the end of appearance in fixed perception, and disdain for art and design. (Original and final participation connect effortlessly to Ulam's seeing *as*, in note 5. They also map separately to Susanne Langer's coeval "presentational" and "discursive" forms of symbolism in *Philosophy in a New Key* (Cambridge, MA: Harvard University Press, 1957). But Ernst Cassirer is a precursor in *The Philosophy of Symbolic Forms* (London and New York: Routledge, 2021), where aptly, to start the first part of *Volume 2: Mythical Thinking*—"objects are not 'given' to consciousness in a rigid, finished state, nakedly in themselves . . . the relation of representation to the object presupposes an independent, spontaneous act of consciousness," 37. Cassirer, of course, has a strong rival in Martin Heidegger. It's fun to try Barfield's terms on both of them—then, Cassirer stands for the rational development of representations in final participation, while Heidegger, it seems, longs for a return to the magic of instinctive consciousness in original participation. More on Langer, and shapes and symbols, is in *Shape*, 18 and 53.) The dimension i for shapes comes up again in subsequent notes, first and foremost in the middle third of note 29. Some of this may seem somewhat repetitious, and it's for a good reason. Judging from my students, dimension is hard to get right—maybe because most of my students are architects, securely tethered to geometrical dimensions in space and ensnared in Cartesian coordinates. Time and again, designs, say, buildings and sculpture, are 0-dimensional in 3-dimensional space. This is the paradigm for building blocks like Lego and their kin, although when I was small, I used to cut wooden blocks into pieces with a coping saw—a jigsaw was too dangerous—and glue these parts together in queer ways. I'd incorporate what I'd done in my designs, painting it first to conceal all the joints—surface and appearance were everything, and at least for me, this is still what goes today. The 0-dimensional worlds of building-blocks weren't mine, with their tedious limitations and their look-alike results—painting added plenty to this, and equally drawing with pencil and paper. But what was only implicit then is explicit now—painting and drawing keep the eye alive, in motion, both inspired and vital. (Drawing is the classical locus for design in architecture. Leon Battista Alberti says it deftly—"All the intent and purpose of [design] lies in finding the correct, infallible way of joining and fitting together those lines and angles which define and enclose the surfaces of the building." In fact, lines and angles are the original stuff of shape grammars—and once fixed, lines and angles define spatial relations. Alberti returns in another guise in note 44. Whether line drawings by hand are welcome or not depends on the designers you happen to ask—I'll let the engineers and the computer experts who buttress their work speak for themselves, soon enough in A3 and in note 27.) And dimensionality contributes a lot more—if painting and drawing aren't 0-dimensional, then just what dimensions are they? Is this always the same? What varies as dimensions do? How do dimensions interact mutually in seeing, and making with different materials? The answers to such questions are in the exuberant (immature/impulsive/spontaneous) eye in original participation, when figuration (embedding) is untamed. Wild things are rampant in childhood and even beyond, everywhere in art and design—inside Max's room in Maurice Sendak's drawings, and outside of visual analogies in von Neumann's pictures and the Rorschach test. In shape grammars, wild things jump out wherever you look, even when you're looking at wild things.

7. Henri Bergson finds something like this in science (physiology) in the groundbreaking methods and practices of Claude Bernard—

> We are in the presence of a man of genius who, having first made great discoveries, inquired afterwards how one should go about making them: an apparently paradoxical sequence of events and yet the only natural one, since the inverse way of proceeding has been tried much more often and has never succeeded.

In fact, this "paradoxical sequence of events" is entirely natural in shape grammars, when rules apply freely in terms of embedding. Things aren't resolved in advance, only in retrospect—once calculating ends. Bergson's remarks are rendered in English in E. Wind, *Art and Anarchy*, 3rd ed. (Evanston, IL: Northwestern University Press, 1985), 140. Wind's take on genius (artistic creation) syncs pretty much with Bergson—ideally, art isn't forced ("*voulu*"), never decided in advance or otherwise prescribed, for artist and audience alike, 77. However with naught to add to this, Wind settles for the patron and program in the "forecourt" of art and design as the impulse for creative activity, 78. Of course, there's more than this kind of "functionalism" to motivate genius and to keep it on track—the full taxonomy also includes "rationalism" when making and materials are key, and "positivism" when functionalism and rationalism interact mutually. (Engineers and the computer savvy may find rationalism a little quaint given current technology in 3D printing, self-assembly, smart materials, etc. Ongoing advances in AI and machine learning may diminish the luster and lure of functionalism, as well.) That genius is explained afterward seems just about right—the influence of both functionalism and rationalism is scant to start, indeterminate if not irrelevant, whether the two work together or not. That real genius is an outcome of rules and embedding is a surprise—nifty to some and less so to others. (The matching passage in H. Bergson, *The Creative Mind: An Introduction to Metaphysics*, trans. M. L. Andison (Mineola, NY: Dover Publications, Inc., 2007), 170, puts Bernard alongside Descartes to survey the "spirit of invention," swerve after swerve. But credible accounts of creative practice are sporadic and rare— "only twice in the history of modern science . . . to determine the general conditions of scientific discovery. This happy combination of spontaneity and reflection, of science and philosophy, happened both times in France," 170–171. Of course, who would dare suggest anywhere else? Maybe it's the *goût de terroir*.)

8. O. Barfield, *Saving the Appearances*, 146.

9. Barfield isn't reticent about this—

> The appearances will be 'saved' only if, as men approach nearer and nearer to conscious figuration and realize that it is something which may be affected by their choices, the final participation which is thus being thrust upon them is exercised with the profoundest sense of responsibility, with the deepest thankfulness and piety towards the world as it was originally given to them in original participation, and with a full understanding of the momentous process of history, as it brings about the emergence of the one from the other, 147.

10. O. Barfield, *Saving the Appearances*, 45, 147.

11. O. Barfield, *Saving the Appearances*, 43.

12. S. Chermayeff, "Let Us Not Make Shapes: Let Us Solve Problems," in *Four Great Makers of Modern Architecture: Gropius, Le Corbusier, Mies van der Rohe, and Wright* (New York: Da Capo Press, 1970), 259. The excitement and professional pride in Chermayeff's remarks is palpable—a bright new future not to be missed.

13. L. Martin, "Architects' Approach to Architecture," *RIBA Journal* 74 (1967): 191–200, 192. A nice title—in short, it's "AAA." Are letters and their combinations all there is to see? Why are words limited in this way?

14. G. Stiny, "A Note on the Description of Designs," *Environment and Planning B: Planning and Design* 8 (1981): 257–267. And also, G. Stiny, "What Is a Design," *Environment and Planning B: Planning and Design* 17 (1990): 97–103. G. Stiny, "Weights," *Environment and Planning B: Planning and Design* 19 (1992): 413–430.

15. J. Ruskin, *The Seven Lamps of Architecture* (New York: Farrar, Straus, and Giroux, 1981), 16.

16. P. Steadman, "Research in Architecture and Urban Studies at Cambridge in the 1960s and 1970s: What Really Happened," *The Journal of Architecture* 21 (2016): 291–306.

17. C. Alexander, *Notes on the Synthesis of Form* (Cambridge MA: Harvard University Press, 1966); C. Alexander, *A Pattern Language: Towns, Buildings, Construction* (New York: Oxford University Press, 1977); C. Alexander, *The Timeless Way of Building* (New York: Oxford University Press, 1979). Tellingly, Alexander's pattern language is a big hit among software engineers and object-oriented computer programmers. L. March and C. F. Earl, "On Counting Architectural Plans," *Environment and Planning B* 4 (1977): 57–80.

18. I took Marvin Minsky's subject in the spring semester of 1966. The class turned out to be more exciting than anyone expected. There were many famous visitors in AI and pointed back and forth, but things really got heated when Minsky angrily asked Hubert Dreyfus ("Alchemy and Artificial Intelligence," Document Number P-3244, The Rand Corporation, Santa Monica, CA, 1965; and later, *What Computers Can't Do* (New York: Harper and Row, 1972)) to leave, and

Dreyfus did, utterly speechless that Minsky insisted on lecturing only to a sympathetic audience in a closed classroom. Minsky didn't pull his punches—"your arguments are invidious"—and never backed down, but this time, no luck. Dreyfus was right—Soren Kierkegaard proved stronger than Turing, but not shape grammars. In fact, Dreyfus's arguments go just so far; they work against symbolic calculating and "counting out," as they support visual calculating with shapes and rules. For added details of the Minsky-Dreyfus affair, as I like to remember them, see *Shape*, 17, 36, 51. In his later years, Minsky was mentioned in some pretty sordid sex trafficking, along with other stars at the MIT Media Lab, and international celebrities, including Prince Andrew, the Duke of York. I like to think that nasty sex crimes and like perversions are independent of AI and its shortcomings—after all, mind and body are supposed to be separate. But then again, maybe not—doesn't the architectural scientist rape architecture in order to expose naked built form (see p. 14)? This kind of violence isn't necessary in shape grammars. The emphasis shifts to art and design, to nudes and sartorial splendor.

19. G. Stiny, "Kindergarten Grammars: Designing with Froebel's Building Gifts," *Environment and Planning B* 7 (1980): 409–462.

20. The limitations of this approach in shape grammars are discussed in G. Stiny, "Spatial Relations and Grammars," *Environment and Planning B* 9 (1982): 113–114, and for more, see note 29.

21. J. von Neumann, "The General and Logical Theory of Automata," in *Cerebral Mechanisms in Behavior: The Hixon Symposium*, ed. L. A. Jeffress (New York: Hafner Publishing Company, 1967), 23–24.

22. J. McCarthy, "Why SAIL and Not Something Else: The History of AI," in *Celebration of John McCarthy's Accomplishments* (Stanford, CA: SAIL, 2012). This transcribes the first half of a video; in the Q and A that's not transcribed, Donald Knuth dismisses McCarthy—"This isn't a rebuttal, you can't argue aesthetics [seeing]." I guess that's because everyone has his/her own aesthetics, that's a "function of his[/her] whole personality and his[/her] whole previous history," and that may show "what kind of a person he[/she] is." But McCarthy didn't need proof; he already knew that intelligence (argument) had nothing to do with aesthetics, and he told me so (c. 1972). I had yet to do von Neumann and the Rorschach test—surely, there, too, intelligence plays no part. Argument is OK in logic, including computer science and AI, because that's what argument is about—the laws of thought work in a deliberate way. However without beauty and delight to support truth, AI seems doomed to fail. It's right, undeniably so, that aesthetics is separate from logic in common practice—you're an artist over the weekend and a scientist hard at work Monday through Friday. But keeping to the status quo is way too easy, and in fact, there's a definite relationship in which one implies the other. And it's not what you think—it's what you see. Logic (symbolic calculating) is a special case of aesthetics (visual calculating).

23. F. Reuleaux, *The Kinematics of Machinery*, trans. and ed. A. B. W. Kennedy (New York: Dover Publications, Inc., 1963), 87–92. This isn't only for engineering kinds of problems; the method works perfectly in art, too. Long before Reuleaux, William Hogarth used a specimen of the "serpentine" as the mark of variety in a parametric scheme for beauty and grace—

VARIETY

In *The Analysis of Beauty* (London: J. Reeves, 1753), 38, he notes "That curved lines as they can be varied in their degrees of curvature as well as in their lengths, begin on that account to be ornamental." Variables (parameters) like these, say, amplitude/height and length, with values greater than 0, define a visual analogy for the serpentine in its symmetrical forms. But consider the limits of these ranges of choice—if amplitude alone is 0, then the serpentine is a flat (straight) line of given length; alternatively, if length is 0, there's an orthogonal line of given height; and for both at 0, a dimensionless point. These singularities are too stern for beauty (ornament) and grace.

24. For a comprehensive review of BIM, see C. Eastman, et al., *BIM Handbook: A Guide to Building Information Modeling for Owners, Managers, Designers, Engineers and Contractors*, 3rd ed. (Hoboken, NJ: Wiley, 2018). The absence of architects and artists in the title is conspicuous—they go way beyond BIM, and owners, managers, designers, engineers, and contractors to appearance. This is exactly what von Neumann worries about, and what Oscar Wilde embraces wholeheartedly—a little father on, near the end of A3.

25. The recursive language LISP ("LISt Processor") with its endless lists is John McCarthy's lasting accomplishment—it's a truly neat way to write code for the structure in descriptions and representations. J. McCarthy, *LISP 1.5 Programmer's Manual*, 2nd ed. (Cambridge, MA: MIT Press, 1965).

26. R. C. Berwick and N. Chomsky, *Why Only Us: Language and Evolution* (Cambridge, MA: MIT Press, 2016). Merge without tampering is a moving target. As science in progress, Chomsky's linguistics is hard to pin down, but isn't this ambiguity/dexterity as much aesthetics/art as logic/science—similar shapes viewed at different times? Of course, some balk at art and design. Nonetheless, shape grammars and generative linguistics align nicely—shapes and rules vs words and rules. And goals match, to trace limits ("competence") minus prediction ("performance"). Although visually incomplete, parametric design may also be the same.

27. I. Sutherland, "Structure in Drawings and The Hidden-Surface Problem," in *Reflections on Computer Aids to Design and Architecture*, ed. N. Negroponte (New York: Petrocelli/Charter, 1975), 75. Sutherland is remarkably clear about the use of structure in CAD, and seems totally uninterested in (oblivious to) how you might calculate with drawings that have no structure, and why this might be key in art and design—

To a large extent it has turned out that the usefulness of computer drawings is precisely their structured nature and that this structured nature is precisely the difficulty in making them. . . . An ordinary draftsman is unconcerned with the structure of his drawing material. Pen and ink or pencil and paper have no inherent structure. They only make dirty marks on paper. . . . The behavior of the computer-produced drawing, on the other hand, is critically dependent upon the topological and geometric structure built up in the computer memory.

The structure of computer drawings is the reason they're hard to make, while the structureless drawings of the "ordinary draftsman" are easy. I guess it's natural to confuse being hard with being important—that's the lure of structure. Of course, drawing (calculating with shapes) invariably supersedes structure (visual analogies). That there's no reason to give up on the artist's way of drawing in order to calculate would have probably been a big surprise to Sutherland, and to others in CAD, when Sketchpad was invented—but no respectable engineer thought much of the artist's methods anyway. Today, as well, this goes for CAD and parametric modeling—little, if anything, has changed when it comes to calculating and the seemingly pathological need for structure. (I first tried the opposing relationship between Sutherland's Sketchpad and my shape grammars in "What Designers Do that Computers Should," in *The Electronic Design Studio: Architectural Education in the Computer Era*, eds. M. McCulloch, W. J. Mitchell, and P. Purcell (Cambridge, MA: MIT Press, 1990), 17–30. And I returned to this relationship in *Shape*, see especially 133–135, 156.)

28. H. A. Simon, *The Sciences of the Artificial*, 2nd ed. (Cambridge, MA: MIT Press, 1981), 185–190. That design is like painting bears the influence of James G. March, who like Simon dreams of "combinatoric design"—the reciprocity is conspicuous. But one good idea needn't lead to another. Creativity in art and design isn't like Lego; it's more than building blocks and units that are given in advance, ready to combine. For many, it's impossible to give up counting for seeing. What would scientists, engineers, and the masters of business and finance say if I asked them to try? Combinatorial design isn't the kind of dream I enjoy having, although sometimes it drifts into pleasant reverie when I compare it to creative design in the analogy that sums up my preface—counting is merely the lower limit of painting, that seeing exceeds effortlessly when there's embedding in shape grammars.

29. James Gips and I worked together on hand tools, as undergraduates at MIT. We were able to describe them as configurations in finite/regular expressions. In retrospect, this was our start on shape grammars, and, in fact, just a few years later (1972), we published our seminal essay. It's worth noting that originally shape grammars were paired with material specifications, including coloring rules and like devices for painting and sculpture. In fact, there's always been a close relationship between shape grammars and making—formatively, 40 years ago in the rational approach to Chinese ice-ray lattice designs (G. Stiny, "Ice Ray: A Note on the Generation of Chinese Lattice Designs," *Environment and Planning B* 4 (1977): 89–98) and in the thumb and forefinger rule in "Kindergarten Grammars" to define and use spatial relations. This is mostly craft and handwork—not too surprising, as shapes themselves are drawings. An early and influential use of color in shape grammars is in T. W. Knight, "Color Grammars: Designing with Lines and Colors," *Environment and Planning B: Planning and Design* 16 (1989): 417–419. Gips did the first computer implementation of shape grammars with applications in artificial perception, as a PhD student

at Stanford in the AI Lab (SAIL)—he readily deferred to McCarthy, and his catchy slogan. In fact, Gips turned to visual analogies to represent shapes, hewing to standard computer practice, then and now. But even with these infelicities, his original software showed the usefulness of visual calculating and opened a key area of research that continues to thrive. At the same time, I was doing my PhD at UCLA. I focused largely on embedding and how shapes fuse when rules are tried in shape grammars (reduction rules for maximal elements, registration marks—"points of intersection"—for partitions of shapes and their transformations, etc.), and on reciprocal relationships between visual calculating, and art and design. Building on the formal part of this work, Ramesh Krishnamurti did the first computer implementation of shape grammars to take embedding seriously, as the starting point for calculating—this was somewhere around 1980, the first third of the decade. Gips and I published our essay on shape grammars before we started our PhD studies at Stanford and UCLA—G. Stiny and J. Gips, "Shape Grammars and the Generative Specification of Painting and Sculpture," in *Information Processing 71*, ed. C. V. Freiman (Amsterdam: North-Holland, 1972), 1460–1465. Our PhD dissertations followed a few years later—J. Gips, *Shape Grammars and their Uses* (Basel: Birkhäuser, 1975) and G. Stiny, *Pictorial and Formal Aspects of Shape and Shape Grammars* (Basel: Birkhäuser, 1975).

Symbols with visual analogies and spatial relations aren't shapes; they differ in terms of the dimension i = 0, 1, 2, 3 . . . of basic elements—points, lines, planes, solids, etc. For i = 0 (points), things are like symbols, where embedding is simply the special case of identity; otherwise, things are like shapes, where embedding is expressed fully and shapes have indefinitely many different parts. Gips's computer implementation of shape grammars was shamelessly 0-dimensional, with neither ambition for nor awareness of anything besides visual analogies—later, he would insist that this was all computers could do, and pooh-pooh embedding. In contrast, Krishnamurti made computers that were 1-dimensional throughout, and that did a lot more. Going from 0 to 1 may seem easy enough—merely a step—but don't be fooled, it marks the enormous shift from counting to seeing and ambiguity, and a pronounced change in what's required to calculate. This subsumes Samuel Taylor Coleridge's prescient division that separates "fancy" (i = 0) and "imagination" (i > 0). The split is described in greater detail near the end of A4 and implicitly, in the preceding parentheses on Norbert Wiener's distinction for counting and statistics in *Cybernetics* (see note 59 and the concluding lines in note 65), and how this overlaps with identity and embedding in shape grammars. Fancy and imagination also run throughout my ensuing answers. Like many other things, fancy and imagination go together (i ≥ 0), so that both are kinds of calculating that meld for mutual advantage. (Whether or not Coleridge's imagination extends this far is an open question.) The first time I considered calculating in contrasting ways was for shape grammars and set grammars, where the latter are symbolic and always 0-dimensional—the members of sets (shapes in spatial relations) are distinct and independent, in the same way points are when they combine. This isn't so for lines, etc. that fuse according to reduction rules—and these reduction rules work for embedding, as well. My original account of set grammars is short and gets right to the point in "Spatial Relations and Grammars."

In tandem with shape grammars, Gips and I outlined an algorithmic approach to aesthetics, highlighting "constructive" and "evocative" modes of description ("interpretation"), and their composition in "aesthetic systems" for design and criticism. Forty years later, this seems mostly trivial stuff, at least to me. Constructive and evocative descriptions are indistinguishable in shape grammars—rules for one work for the other in exactly the same way. Identities x → x provide good examples for both; they construct what we see, as they bring in evocative names and words. This may not satisfy everyday generative/constructive expectations of adding simple shapes together to get more complicated ones, but it's constructive nonetheless—identities can make simple shapes, starting with a lonely line, pretty complicated, with plenty of internal articulation and structure, etc. (For example, see my "Useless Rules" in note 3.) And to extend this, every rule in the embed-fuse cycle is its own little "design algorithm" and "criticism algorithm," including "receptors" and "effectors"—after all, rules are for seeing and doing. But algorithmic aesthetics won't go away; there's still more than ample enthusiasm for it, especially for the evaluation measure $E_z$ that applies to compare unity (in sync with Ockham's Razor) and variety, or order and complexity. Everyone enjoys counting units when things are already divided up, but who knows, there may be alternative ways to make this right—for aesthetic (evaluation) measures without fixed parts and units. If nothing else, Coleridge is keen on reducing multitude (variety) to unity in a kind of poetic (natural/organic/romantic) process in which things fuse in undivided (largest) wholes, in order to re-divide and re-create (see the parentheses that concludes A3, and the discussion of imagination in A4); as engineers, Gips and I framed $E_z$ mathematically, to incorporate algorithmic information theory and relative entropy, counting discrete symbols or smallest units in a standard, predefined vocabulary—tediously, unit by unit. But no more—I've given up on technical/mechanical devices. Today, my sympathies are entirely with Coleridge and his organic analogues. The fluidity of perception implies that aesthetic value can vary—how this "adds" up in a single picture or poem needs to be worked out. I've always been something of a closet romantic, seeking delight in everything I see—not forever, but at least for the time being. (Maybe this essay is my way of coming out.) And of course, there's some neat math in all of this, too. Visual calculating in shape grammars, when there's embedding, and shapes and their parts fuse, assimilates algorithmic aesthetics without loss—G. Stiny and J. Gips, *Algorithmic Aesthetics: Computer Models for Criticism and Design in the Arts* (Berkeley, CA: University of California Press, 1978). (After many years with an exclusive focus on shape grammars, I started to teach algorithmic aesthetics again in my graduate seminar. The big surprise was how accessible my students seemed to find it. They assimilated aesthetic systems in computer models effortlessly and had plenty to say about how they should work, while shape grammars and the embed-fuse cycle took time, lots of it, mostly in confused silence. I'm forever amazed when thinking is easier than seeing—when visual analogies are enough for everyone to take in everything. I guess that's the result of too much training, and rote recitation. Is it any wonder that many people worry that the growing reach of computers and AI will soon overtake every aspect of life, when all they do is think? The ambiguity—grammatical anaphora—cuts both ways, for people who only think by choice and for computers that do by definition, so that who and that aren't distinct. There's more on the reach of computers and AI in the coda at the end of A7.)

30. A helpful summary of CAD at MIT is in D. Cardoso Llach, "Software Comes to Matter: Toward a Material History of Computational Design," *Design Issues* 31 (Summer 2015): 41–54. It seems that Douglas Ross had high hopes for the "plex" as a "scientific" way to describe/represent anything, simply in definitions. The transition from given data to structure to algorithms is worth trying, to see how far it goes. But maybe some things are ineffable; you can never describe/

represent them in full, and once and for all. They aren't eternal, as philosophers like to say. No matter how hard you try, there's something that's missed—every description is incomplete. Isn't this what von Neumann worries about in the Rorschach test—that it's beyond the scope of visual analogies? This is certain for von Neumann, and clinically decisive, as well. Maybe it's better to hold your tongue and not say too much, in order to forgo the asylum and a padded cell. Keep what you see to yourself—presto, it's bound to alter in a delightful way. Flouting recognized visual analogies (collective representations) is too perilous for free rein in a civilized (polite) society.

31. S. A. Coons, "An Outline of the Requirements for a Computer-Aided Design System," *AFIPS'63: Proceedings of the Spring Joint Computer Conference*, Detroit, Michigan, May 21–23, 1963, 299–304.

32. C. M. Eastman, "Information and Databases in Design: The Computer as a Design Medium," in *Representation and Architecture*, eds. Ö. Akin and E. F. Weinel (Silver Spring, MD: Information Dynamics, 1982), 245.

33. For how description rules are defined and apply, see note 14—calculating in this way subsumes Coons's "design process." Other ways to describe shapes in terms of topologies is spelled out in note 3.

34. W. James, *The Principles of Psychology* (Cambridge, MA: Harvard University Press, 1981), 956–958. James is also compelling a page later, in fact, throughout *The Principles*—

> All ways of conceiving a concrete fact, if they are true ways at all, are equally true ways. *There is no property* ABSOLUTELY *essential to any one thing.* The same property which figures as the essence of a thing on one occasion becomes a very inessential feature upon another, 959.

I guess this is another way of saving the appearances—so much the worse for Ockham's Razor, in James's easy embrace of Epicurean plurality. Ockham seems to be mostly an inadequate judge of truth or at least incomplete, and says scant about art and design. Why is his razor so sharp for science, and computers and AI? It seems it's a lot easier to count than to see—it's more trustworthy and sometimes, it's repeatable, too. My discussion of sagacity and embedding is in *Shape*, 64–65.

35. G. Stiny, "*The Critic as Artist*: Oscar Wilde's Prolegomena to Shape Grammars," *Nexus Network Journal: Architecture and Mathematics* (2016): 1–36.

36. O. Wilde, "The Critic as Artist," *Intentions*, in *The Artist as Critic: The Critical Writings of Oscar Wilde*, ed. R. Ellmann (Chicago: University of Chicago Press, 1982), 369.

37. Wilde's critical spirit seems especially right for art and design—inasmuch as there are indefinitely many ways to see anything as in itself it really is not, and there's just one way to see it as in itself it really is. On the one hand, visual analogies are incomplete, with no way for anyone to fill in all the gaps, even for him/herself; and on the other hand, visual analogies make new perception unnecessary—they're true and the final substitute for what there is to see. Of course, there may be some things (pictures and poems, shapes and figures of speech, figures of other kinds, Hamlet and you and me, etc.) for which there's no way to see them as in themselves they really are. Maybe then, they're the succession of what as in themselves they really are not. Wilde's

beautiful form and von Neumann's Rorschach test intersect in a splendid asymmetry—visual calculating in shape grammars is inevitable, first for visual analogies (trivially for the dimension $i = 0$, see note 29) and then for what as in themselves visual analogies really are not (for dimension $i > 0$). Aesthetic (sensory/perceptual) experience is inherently open-ended in art and design—what would life be like without it? Is it any wonder that aesthetics trumps logic and ethics? Ambiguity is the quick of shape grammars.

38. Examples abound to show how art misconstrues science, and vice versa. In the first installment of his 2016 "Stanislaw Ulam Memorial Lecture" at the Santa Fe Institute, the MIT engineer/scientist Seth Lloyd described his fabulous visit to the *Biblioteca Casanatense* in Rome, to introduce the wonderful conceit he had written out in red (bold), blue (italics), and black ink, for everyone to see—

> A physical law is **like** a metaphor, but it is **not** a metaphor.
> Physical laws are *strictly more powerful* than metaphors.

But questions from the audience were on quantum effects and relativity—what about the physics of metaphor? The reason Lloyd's metaphor fails, and for his conclusion, too, is that "physical laws are mathematical and descriptive, and they are [numerically] predictive." Not terribly convincing—solutions to physical laws, whether you can find them or not, come entirely at the expense of ambiguity and so metaphor itself. And what happens to creativity without ambiguity and metaphor—is science any different than art in this regard, in the way new phenomena and known are observed and described? In neither art nor science is lawful certainty a productive substitute for lawless (willful) seeing. Is it any wonder that art and science continue to mistrust one another? Diminishing art (ambiguity) to prove the superiority of science (predictability) is no way to bridge this divide; in fact, it doesn't pay to value science more than art—ignoring ambiguity hobbles both indiscriminately. In shape grammars, the inclusion goes the other way around, without any loss to either art or science—ambiguity in metaphor (a beautiful form) and the Rorschach test assimilates the predictive power in formulas and physical laws, visual analogies, etc. I plot the relationship between perception (ambiguity) and prediction in the preface, as well, after Alberti's portrait medal. There are some added bits on observation and art in Alberti's *On Sculpture* in note 44. Somehow, it seems right that the inventions of metaphor aren't entirely understood; two alternative accounts are in the parentheses on pp. 40–41—one relies on figure-ground reversal (ambiguity) in Gestalt psychology to debunk the rigorous mappings (formulas) invoked in the other. This leaves the wiles of creativity for the preface, p. 14, p. 23, the beginning of the parentheses from p. 33 to p. 40, and pp. 47–48, and notes 7, 28, 47, 64, 65, 68, and 72; and how physical laws deaden experience for note 77. Words in descriptions/metaphors are the unforeseen culprits in note 77; formulas and numbers are duplicitous in the same way. To be open and above board, I must also say that up until recently, Lloyd shared some of Minsky's unsavory friends (see note 18). Maybe Lloyd's formula for friends failed in a prediction error from priors, providing new data for science—does it matter? The official (party) line at MIT is that even an engineer/scientist can make "a mistake in judgment." I hope so—isn't this the locus of metaphor, and a metaphor for what metaphor is? Judgment and the feelings of sentiment are compared in note 70; the sway of "modern judgment" is traced in the preface.

39. M. H. Abrams, *The Mirror and the Lamp: Romantic Theory and the Critical Tradition* (Oxford: Oxford University Press, 1971), 174.

40. Reduction rules are key to change descriptions and visual analogies (sets) into shapes. Otherwise, the synthesis of a thesis and antithesis wouldn't be "indifferent"—it would still favor fixed parts, for example, squares and triangles or a list of common parts for both. One way to define common parts that shows how they're related is given in example 1 in Exhibit 2. For the original definition of reduction rules in shape grammars see *Pictorial and Formal Aspects of Shape and Shape Grammars*, 152–161. Reduction rules $R^+$ for synthesis are given for lines in table 3 in Exhibit 1.

41. Reflection often follows untutored (raw) experience to turn it into thought, in fixed descriptions (visual analogies) and formulas that add to a cache of surefire practices, ready for automatic use by the professional architect and designer. It's not unusual for this to take away the looking—maybe that's the whole point of it, to make new perception unnecessary. Isn't this the goal of computers and AI? Shape grammars put the looking back, entirely in terms of the embed-fuse cycle.

42. The first time I used schemas explicitly was in "Kindergarten Grammars"—a spatial relation for the shapes A and B defines the schema $x \rightarrow A + B$ and its inverse $A + B \rightarrow x$, where the variable x is either A or B. Typically, A and B are modified ("labeled" with points and sometimes symbols) in terms of their symmetries, to control how rules apply in varied ways. Rules are tried recursively to derive shapes of many distinct kinds—it takes surprisingly little for a rich repertoire of designs, when spatial relations are used to calculate, and then kept to. Nonetheless, no combinatory program like this seems to be enough for full-blooded design. The excess in looking is what makes design so compelling.

43. Marcel Duchamp and his friends, Beatrice Wood and Henri-Pierre Roché, are particularly good on this in the May, 1917 issue of *The Blind Man* in The International DADA Archive (Iowa City IA: The University of Iowa Libraries). In an unsigned editorial in the second and, alas, final issue of their journal, they describe R. Mutt's *Fountain* as "plagiarism, a plain piece of plumbing." Mutt's theft is trivial and "unique, utterly different"—

> Whether Mr. Mutt with his own hands made the fountain or not has no importance. He CHOSE it. He took an ordinary article of life, placed it so that its useful significance disappeared under the new title and point of view—created a new thought for that object.

Mutt is, in fact, the ideal blind man—unable/unwilling to see in the expected way, in order to see in another. But how is this possible without rules, and embedding and shapes that fuse? (This is neatly related to Wilde's critical spirit expressed in the aesthetic formula to see things as in themselves they really are not, and also to Salvador Dali's "paranoiac magic" and "negative hallucination" in note 66.)

44. L. B. Alberti, *On Painting and Sculpture*, trans. and ed. C. Grayson (London: Phaidon, 1972), 121. Alberti's remarks at the beginning of *On Sculpture* are worth reading in full. With embedding in shape grammars, they extend easily from "a tree-trunk or clod of earth" to von Neuman's

Rorschach test and Wilde's beautiful form, and to painting and sculpture, and to whatever anyone sees. This explains art and design in terms of observation; everything depends on seeing—

> [Artists] probably occasionally observed in a tree-trunk or clod of earth and other similar inanimate objects certain outlines in which, with slight alterations something very similar to the real faces of Nature was represented. They began, therefore, by diligently observing and studying such things, to try to see whether they could not add, take away or otherwise supply whatever seemed lacking to effect and complete the true likeness.

Art and design are for the eye and not the mind; they go beyond visual analogies. Nonetheless, this seems hard to keep—Alberti opts for the mind in *On the Art of Building in Ten Books*, trans. J. Rykwert, N. Leach, and R. Tavernor (Cambridge, MA: MIT Press, 1988), 7—

> It is quite possible to project whole forms in the mind . . . Since that is the case, let lineaments be the precise and correct outline, conceived in the mind, made up of lines and angles [spatial relations], and perfected in the learned intellect and imagination.

Maybe so, but five and a half centuries later, the eye still reigns supreme. The abstract painter Ellsworth Kelly is explicit about this—"In my paintings I'm not inventing; my ideas come from constantly investigating [observing] how things look." And things may equally include Kelly's own paintings and what other artists do—paintings/designs rely on other paintings/designs. These kinds of fluid relationships provide the basis for art and art criticism; they're easy to frame in shape grammars. I can approach this constructively, generating new designs from scratch in a certain style, or analytically, copying a known design or parti to get new designs—"reading" them off according to different identities x → x. For example, try the parti

with triangles and K's to get these two remarkably different designs (follies)

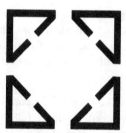 

(Uninformatively, both designs are the same built form.) One is a dungeon in heavy brick, anchored at corner prison cells, while the other is an airy pavilion in glass and steel, supported with minimal brackets in its side walls—really slick. And both of these designs on paper are ready for contrasting anthropomorphic (evocative) descriptions—menacing folded arms and welcoming open ones, arrows blunted in closed inside spaces and pointed expansively outward. I used to emphasize constructive methods more than analytical ones; now, it's both equally, in reciprocal processes that rely on seeing. I first traced constructive and analytical methods more than forty years ago, in "Two Exercises in Formal Composition," *Environment and Planning B* 3 (1976): 187–210. This is where I introduced spatial relations (I used them later in "Kindergarten Grammars") as an original way to describe shapes and to define rules in shape grammars, but I didn't know then that spatial relations connected up so neatly with visual analogies, and to von Neumann and McCarthy. I'm still surprised by how much "Two Exercises" includes. I'm inclined to think I had an uncanny streak of crazy luck—in many ways, this was the beginning of shape grammars as a viable research program in art and design that's livelier than ever, and even now, pretty unique.

45. "Transmemberment" is in Hart Crane's "Voyages, III"—one of those wonderful words that sounds right and feels right instantly, with no easy definition. Going from Crane's "song" to shapes shows how it works in the embed-fuse cycle in shape grammars—at the quick of experience, where shapes are changed and reordered in terms of parts and relationships, charged with meaning in new perception. Whether this is immediate in the flash of a single rule application or plays out by fits and starts in multiple steps doesn't matter—as I've already shown for any rule A → B, and parts of A and parts of B, every rule application is many and many are one in the embed-fuse cycle. I guess the direct connection to shape grammars and visual calculating is the real reason I like the word so much. Sometimes, however, I try to think about transmemberment in more conventional terms—for sets (spatial relations) and their members that fuse and re-divide to define equivalent sets that supersede what's already given, even if this relies on willful tampering and encourages it. (It's the same for separate pieces of clay that are pressed into a single lump and cut into pieces again, although in general, parts needn't be discrete.) This brings in Coleridge's synthesis again, to link thesis and antithesis as feeling swells and breaks in waves, in an inconstant array of untallied possibility. But isn't this shape grammars? It's amazing how many varied ways there are to change things, and how words invariably fail to describe this process fully—transcribing, transfiguring, transforming, transgressing, transmuting, transposing, transvaluing, each is no more than a faint and tentative start. Such lexical similarities inform shapes and shape grammars, yet none matches the poet's "silken skilled transmemberment of song"—visual calculating with rules in the embed-fuse cycle outstrips common words in common use. It usually goes unnoticed—that shapes and song (poems) are related reciprocally, at the very least to exemplify and describe. And there are technical means for this even now, in description rules and retrospective topologies—for a few details, see note 3 and note 14.

46. M. Garber, *The Use and Abuse of Literature* (New York: Pantheon Books, 2011), 238, 253–255, 258.

47. This runs from poems to ordinary language—à la Chomsky, where words (atoms) combine according to inbuilt rules in a generative (recursive) process that's finite with infinitely many

(unbounded) results. The words we choose somehow fit the vagaries of place and time, influenced by these coordinates yet not determined/dictated by them—it's a creative trick and a total mystery. Still, the odds of success seem short when ambiguity is everywhere; whatever words we use can't help but fit when there are so many ways to pick out things for the time being—maybe like Alberti in the Quattrocento, diligently observing, adding, and taking away to complete a true likeness/world. (Alberti tries identities $x \rightarrow x$ in the embed-fuse cycle, and rules in the primary schema for parts $x \rightarrow \mathrm{prt}(x)$ and its inverse $x \rightarrow \mathrm{prt}^{-1}(x)$, where $x \rightarrow x$ is the mutual subset of $x \rightarrow \mathrm{prt}(x)$ and $x \rightarrow \mathrm{prt}^{-1}(x)$.) I guess that's why we're so skilled at using words—without being taught. Education, training, and inculcated habits keep us from what we know intuitively, and from what we see to take away the looking. Chomsky ties Coleridge to Wilhelm von Humboldt to show that language is generative—maybe "the infinite I AM" in "the finite mind." This ducks Coleridge's thesis/antithesis/synthesis, and imagination vs fancy—in the analogy with shapes and symbols in the preface and note 29. Chomsky's words and rules are fancy in a generative mode, with "counters" and "fixities and definites." (AI/BIM is the same, from McCarthy's parametric descriptions to machine learning and training sets.) Is imagination generative, too? Chomsky defers to Juan Huarte, the 16th century Spanish physician whose taxonomy of "wit" lists imagination in "a mixture of Madness" after memory and understanding (language)—"Wit is a generative power." Coleridge's imagination or "esemplastic power" is boundless in the embed-fuse cycle, in pictures and poems and across art and design; calculating without counters (atoms) overtakes understanding in the hierarchical structures of language defined by merge and no tampering, and in compositional/combinatorial structures of all kinds. Imagination is generative (creative) beyond understanding. (Chomsky also relies on A. W. Schlegel to put language before imagination in art and design, but this limits the embed-fuse cycle.) See A3 at p. 20, what's next in A4, including notes 49–66, and in A6, note 68. Coleridge, Huarte, Schlegel, and mostly Humboldt are in N. Chomsky, *Cartesian Linguistics* (New York and London: Harper & Row, 1966).

48. O. Wilde, "The Critic as Artist," *Intentions*, 368.

49. S. T. Coleridge, *Biographia Literaria*, eds. J. Engell and W. J. Bate (Princeton: Princeton University Press, 1984), Vol. 1, 304. I consider Coleridge's twin versions of imagination strictly in the embed-fuse cycle, to gloss his unique (historically, cryptic) gloss of Immanuel Kant's critical philosophy. The "primary" imagination is God's sphere, to fill in everyday perception with what's conceived in his/her mind; this is automatic and none of my concern, something I take for granted. Poetic (aesthetic) imagination is "secondary" and syncs with the conscious will; this sparks genius to re-create personal experience throughout art and design, in pictures, poems, etc., and concerns me fully—in the endless go of rules in the embed-fuse cycle, to align with God and maybe not. Computers and AI track toward God, but so far with spotty results; they skirt personal experience, overlooking new perception—hence, the ineluctable need for shape grammars to add looking. In the next long parentheses in A4, Wiener contrasts God and cybernetics in a children's song that's good for shape grammars, too—to complete a trinity in clouds.

50. I. A. Richards, *Coleridge on Imagination* (Bloomington, IN: Indiana University Press, 1960), 75. Richards describes six types of imagination in *Principles of Literary Criticism* (London: Routledge, 2001), 224–236. He starts out with the everyday definition in terms of "vivid images, usually

visual images" in the mind—this is the definition that always comes up when you Google "imagination"—progresses penultimately, to "scientific imagination" in which value (purpose) plays a key role, and concludes in triumph, with Coleridge. Neatly, Barfield in *Poetic Diction* (Middletown, CT: Wesleyan University Press, 1973), 177, distinguishes simply between wonder and strangeness. There's surprise in both, but it's beyond our understanding in the first, and grasped only in "aesthetic imagination" in the second—only when esemplastic power kicks in, "in order to re-create." "Strangeness, in fact, arouses wonder when we do not understand; aesthetic imagination when we do." Calculating with shapes and rules in the embed-fuse cycle is surprising and strange time and again. It's a delight to go back to pictures and poems with a queer eye and ear—to stale unknowns. Nothing is settled, as long as there's imagination. And imagination in the embed-fuse cycle is rife throughout *Poetic Diction*, 32—

> [David] Hume taught us that the world of things is, in fact, a habit of mind. [Bertrand] Russell affirming that 'whatever can be known can be known by science,' denies *a priori* the possibility of disturbing the habit. Reflection on the poetic activity [seeing] teaches us that the same imagination which created that kind of habit can both disturb it and create new ones.

I wonder what this means for final participation—Barfield's "poetic activity" is all visual calculating in shape grammars, to establish new habits retrospectively. Who would have ever thought that calculating, and pictures and poems were how to augment the world of things? Several years ago, I gave a talk to a bunch of philosophers (SPT 2017) who wanted to know about the grammar of things; they were surprised that "grammar" was more than a word, but a tried and true method for calculating new things (shapes), not once, but yes, time after time—so much the worse for Russell, who turns Hume's habits into facts in a scientific world that's stable, with scant for open-ended personal experience, forever inchoate and variable. Visual calculating in shape grammars—again, rules (seeing and doing) in the embed-fuse cycle—makes any habit (rule) easy to break with another habit (rule) that's just as easy to form. To a large extent, we need to be faithful creatures of habit in our daily routines, when useful things concern us the most and make a real difference. There are "common sense" conventions for ordinary activities that need to be followed, for example, crossing Massachusetts Avenue during rush hour, without life-threatening injury. Nonetheless, this isn't everything all the time—there's limitless freedom in ambiguity, and pictures and poems, and in all of the useless things that don't matter in a practical way. Poetic activity, that is to say, visual calculating in shape grammars, is vital in art and design. (Russell speaks for himself in *A History of Western Philosophy* (New York: Simon and Schuster, 1972); he construes philosophy mainly in terms of the social conditions and political events of the day. Chapter XVIII in Book Three, "The Romantic Movement," exposes the root problem—

> . . . having unfortunately been supplied with funds by the Wedgwoods, he [Coleridge] went to Göttingen and became engulfed in Kant, which did not improve his verse, 679.

Coleridge relies on Kant and others for imagination and esemplastic power in the *Biographia Literaria*. René Wellek lists Coleridge's crimes and thefts in an indictment without exculpatory evidence or facts—"[imagination] dissolves, diffuses, and dissipates, in order to re-create"—in *A History of Modern Criticism: 1750-1950* (New Haven: Yale University Press, 1955), Vol. 2, 151-187. (A century later, Mutt faced similar charges.) Whatever the source, the poet's loss in verse is the critic's gain in reach, in a signal trade-off. Of course, William Blake comes before Coleridge and the romantic movement in full flower. Russell locks Blake and Charles Darwin in age-old strife, in which appearance and beauty confront use—

> The romantic movement is characterized, as a whole, by the substitution of aesthetic for utilitarian standards. The earth-worm is useful, but not beautiful; the tiger is beautiful, but not useful. Darwin (who was not a romantic) praised the earth-worm. Blake praised the tiger, 678.

Today, computers and AI reject aesthetic standards for utilitarian ones in scads of data—the earth-worm holds the future, not the tiger. Yet somehow, the critical spirit in perceptual change evolved to defy practical success in tests of utility and common use. Is the critical spirit idle, or does it influence evolutionary change? Maybe aesthetic standards matter. In fact, Darwin ties them to "sexual selection"—appearance/beauty is a fertility test. Blake's "Tyger" isn't useful and thrives all the same, in the fearful symmetry of a beautiful form—fiery-bright in new perception, re-framed, seized and seen in the agile hand or darting eye. Rules in the embed-fuse cycle alter perception in the artist's formula Eye and Hand to include Blake—with Darwin in tow, as Eye and Hand assimilate Mind in search of useful truths. Russell simplifies the relationship for beauty and use in an analogy—

Blake : Darwin :: tiger : earth-worm

This highlights the contrast between John Ruskin's decoration/ornament and the architectural scientist's naked built form encoded in graphs in order to count, and for greater generality, in my original ratio of shapes and symbols. And the way the embed-fuse cycle unites Blake and Darwin goes, as well; visual calculating in shape grammars incorporates appearance/beauty—and use, indirectly in description rules or directly with weights. It's easy for experience in pictures and poems to offset millenniums of philosophy—especially in shape grammars. I should add that Coleridge and Russell embrace the big book; in words, weight, length, and girth, Russell's *History* rivals Coleridge's *Biographia*. Parity in pages is a measure of matching impact in literary biography and intellectual history.)

51. I. A. Richards, *Coleridge on Imagination*, 101. Common use includes at least as much as words (atoms) and merge (recursion) in Chomsky's minimalist account of language—and very likely more.

52. I. A. Richards, *Coleridge on Imagination*, 72.

53. I. A. Richards, *Coleridge on Imagination*, 110. Richards made practical criticism into a discipline, but Coleridge seems to be the original source of the phrase.

54. I. A. Richards, *Coleridge on Imagination*, 92.

55. W. Köhler, "*Cybernetics of Control and Information in the Animal and the Machine* (Review of Book)," *Social Research* 18 (1951): 125–130, 128.

56. W. S. McCulloch and W. H. Pitts, "A Logical Calculus of the Ideas Immanent in Nervous Activity," in W. S. McCulloch, *Embodiments of Mind* (Cambridge, MA: MIT Press, 1965), 19–39. McCulloch was pleased when von Neumann used this essay to teach the general theory of digital computers and automata—W. S. McCulloch, "Recollections of the Many Sources of Cybernetics," *ASC Forum*, VI, 2 (Summer, 1974), 11. I'm glad it wasn't my textbook.

57. L. Hardesty, "Explained: Neural Networks," *MIT News* (April 14, 2017). Keeping abreast of fast moving developments in AI and neural networks is nigh on impossible—major breakthroughs are reported daily. In fact, highflying theoretical physics seems to be next in line to yield to AI and neural networks. "The Theory of Everything is still not in sight, but with computers taking over many of the chores [sic] in life—translating languages, recognizing faces, driving cars, recommending whom to date—it is not so crazy to imagine them taking over from the Hawkings and the Einsteins of the world. . . . [The] tool in this endeavor is a brand of artificial intelligence known as neural networking."—D. Overbye, "Can a Computer Devise a Theory of Everything?," The New York Times (November 28, 2020). In the quote in note 5, Ulam makes the case for seeing *as*, as a prerequisite for AI, and in the rest of the passage (not quoted), he looks for some future Einstein to formalize "as" in logic, like other Boolean connectives. This closes the circle in a convincing and unexpected way—neural networks in AI have the amazing power to turn AI into AI, going from AI (neural networks) to Einstein to seeing *as* to AI (neural networks). Really though, Overbye's list exceeds the narrow bounds of AI today, whatever its provenance in mind and machine, and prior data and facts; in particular, translating languages and recognizing faces are at the quick of art and design, where Coleridge's imagination in the embed-fuse cycle is vital in Wilde's beautiful form and von Neumann's Rorschach test—to judge the depravities in the unsettled picture of Dorian Gray, or to see Picasso in Velázquez and Rembrandt. To be sure, all the computer hype will feed on itself and continue to grow, but what a slight kind of ambition— don't count on AI and neural networks to take over from Einstein or to formalize seeing as anytime soon. Sixty years ago, McCulloch was also uncertain about insight, intuition, and invention, and their fate in AI and neural networks—in the main text after this note. Insight and intuition sway perception (seeing *as*, etc.), even as neural networks grapple with them, in analogies couched in mappings between "vectors" (descriptions/representations). "'Similarities [mappings] of big vectors explain how neural networks do intuitive analogical reasoning' . . . intuition captures that ineffable way a human brain generates insight." Of course, such mappings have already been tried for metaphor, and as Garber shows, with palpable impatience, they take away the looking, fixed throughout and dead. Neural networks are blind to the embed-fuse cycle—it's hard for a neural network "to take on a new structure—a parse tree—for each new image it sees," and presumably, this is the same looking at a single image time and again in new ways. There's neither ambiguity in neural networks nor room for a beautiful form—vectors "echo" one another in monotonous lockstep, fixed terms in a resonant series to see things as in themselves they really are, they really are, they really are . . . . This veers away from insight and intuition in art and design, acutely in pictures and poems. AI is lost in a rote delusion. Siobhan Roberts assays its grim etiology in "Geoffrey Hinton Has a Hunch about What's Next in AI," *MIT Technology Review* (April 16, 2021). Unlike vectors that point in one direction, Hinton's hunch wobbles, although to process and represent information in parse trees in order to show what neural networks do would be an impressive feat, and possibly even to infer images and objects from their parts, a vase from a shard (tree-trunk or clod of earth). Nonetheless—"Science . . . is 'full of things that sound like complete rubbish' but turn out to work remarkably well—for example, neural nets." Maybe so for prior data and facts; maybe so, but follow your eye—and wait and see.

58. W. S. McCulloch, "What is a Number that a Man May Know It, and a Man that He May Know a Number?" in *Embodiments of Mind*, 14.

59. N. Wiener, *Cybernetics: Or Control and Communication in the Animal and the Machine*, 2nd ed. (Cambridge, MA: MIT Press, 1961), 30. The first time I recited this song, I said to myself—"Wow, this is totally fantastic, at last someone else is actually going to try my kind of embedding." However, Wiener dashed my hopes in "statistical dynamics"—sounds like a pop riff. In AI and machine learning today, even with vast and growing neural networks for mappings and vectors, prospects for a real breakthrough haven't improved—there's statistics through and through, with scant notice of "the whole fiery-coloured world," that is to say, calculating in the embed-fuse cycle. Alberti is right that embedding is "diligently observing and studying." But seeing, to see how seeing works, is harder than I thought—maybe it's a little too twisted.

60. I use counting for number in order to sidestep man and number in McCulloch's title, without violating the spirit of God in Wiener's song. This may not matter—number is inherent in counting that also extends to man/woman. Likewise, ambiguity is inherent in seeing— for shapes in clouds and not. I use seeing to counterbalance counting as the preferred way to calculate, so that shapes and numbers are equal. Still, ambiguity in shapes highlights the flux of new perception. Maybe seeing and ambiguity switch—no, this alters meaning. But, poetic diction isn't decisive; soon enough, ambiguity and personality kick in—in metaphor and mistakes in judgment, and in extravagant images in the Rorschach test. Shapes and words make it hard to sort things out for good and all, and with ongoing delight in strangeness and wonder, why try? Ambiguity explains the power of a beautiful form in pictures, poems, and songs; it's an aesthetic alternative to Wiener's two divisions in science—(1) determinism in God and counting, and (2) randomness in cybernetics and statistics. (The triaxial pinwheel in the subgraph in the main text shows how (3) ambiguity in shape grammars and seeing relates to determinism and randomness.) There's boundless joy in half-knowledge, and plenty to use; that's the go of visual calculating—of imagination in the embed-fuse cycle. (Does my aesthetic option fit elsewhere; does ambiguity in the embed-fuse cycle enliven free will beyond determinism and randomness; does it expand consciousness as it strays/swerves from visual analogies? The chances seem slim no matter what I say—ambiguity is ignored outside of art and design, and even with Coleridge and others, ignored too often in art and design, as well. Why look when you know what you'll see? Half-knowledge in artistic/poetic imagination counts for naught—these aren't romantic times.)

61. D. Gondek, *Conversations on Art and Science: Tricksters of Big Data—Artificial Intelligence or Intelligent Artifice*, School of the Art Institute of Chicago, February 19, 2014. David Gondek on Vimeo: www.vimeo.com.

62. D. Sax, *The Revenge of the Analog: Real Things and Why They Matter* (New York: Public Affairs, 2016), 36. I guess I should add cooking after physical fitness in the list that follows Sax, a few lines down on the same page.

63. S. T. Coleridge, *Biographia Literaria*, Vol. 1, 305. The embed-fuse cycle endued with good sense renders Coleridge's fancy and imagination, and their relationships.

64. H. Poincaré, "Mathematical Creation," *The Monist* 20 (1910): 321–335. Bergson (note 7) ignores Poincaré in his remarks on French genius—I guess genius is a high bar, especially in France. It's reassuring to think that this is because Bergson resists the forward thrust of fancy

in Poincaré's account of mathematical creation, to keep instead to the backward push and pull of Descartes and Bernard in philosophy and science—sounds good, but who can really know? Figuring out what others mean is a lot like looking at shapes—I know what you mean when I see what you say.

65. The marvelous figure "[to] think aside" is from the late 19th and early 20th century French philosopher of aesthetics and invention, Paul Souriau. He's quoted in J. Hadamard, *The Psychology of Invention in the Mathematical Field* (New York: Dover Publications, Inc., 1954), 48—"Pour inventer, il faut penser à côté." Paul Valéry's gloss on genius, and combination and choice in poetic invention is recorded a little earlier, 30. Hadamard himself seems obsessively focused on the *"forgetting hypothesis,"* 33, as a crucial aspect of invention—to avoid sliding "insensibly" into a groove or falling into a rut with no means of escape, by "getting rid of false leads and hampering assumptions" that interfere with an open mind, and seeing. The forgetting hypothesis is affirmed every time shapes fuse in the embed-fuse cycle—in fact, every time a rule is tried. No doubt, this makes it the Law of Forgetting, rather than just another hypothesis to test and disprove. In visual calculating in shape grammars, the results of prior divisions (visual analogies) are false and hampering; once they're a habit, when they're rigorously taught and learned, they block open-ended embedding—to see aside/askance—so that there's no chance for change, only rote recitation. This is the end of pictures and poems, that put everything in flux and ignore previous results. Farewell to seeing, and to art and design. Everything stays the same—frozen in memory, lacking vital signs, "essentially fixed and dead." In note 18, I mentioned Minsky and Dreyfus and what AI can and can't do, and how Kierkegaard easily surpasses Turing, to see past past results. Kierkegaard's simple answer to calculating is ideal for symbols, and shapes in the schema $x \rightarrow \Sigma F(prt(x))$ from A4, that includes all rules, each defined in many ways; the passage is in *Kierkegaard's Concluding Unscientific Postscript*, trans. D. F. Swenson and W. Lowrie (Princeton, NJ: Princeton University Press, 1968), 68—

> While objective thought [symbolic calculating] translates everything into results, and helps all mankind to cheat, by copying these off and reciting them by rote, subjective thought [visual calculating] puts everything in process and omits [forgets] the result.

It's fascinating how you can reach the same result (conclusion) in alternative ways, for example, via calculating and pictures, in the Rorschach test and a beautiful form. Aesthetic imagination, visual calculating in shape grammars, is an effective way to find new things in known ones, and to make different things the same—where the new and the same are one in endless relationships that are forever in flux (cf note 37). And this goes for reading poems and writing them, in Coleridge's realm of esemplastic power. At times, I try it in terms of Bloom's six "revisionary ratios," in *The Anxiety of Influence: A Theory of Poetry*, 2nd ed. (New York: Oxford University Press, 1997). By name—*clinamen, tessera, kenosis, daemonization, askesis*, and *apophrades*. But these are at best mere inklings, indistinct outlines and sketches, far from the constructive and analytical methods delineated in shape grammars. Still, it strikes me that Bloom's ratios or something like them are needed to texture imagination for any kind of calculating (re-vision) with poems—enumerating what poets might try in order to go on, in the way I use schemas as heuristics for rules to see and do. In particular, clinamen ("poetic misprision proper") is misreading, or embedding in another way that may seem strange for a while, but that supersedes what's settled and known. (Ratios like Bloom's are appealing to many; Nelson Goodman has a correlative taxonomy for "ways of worldmaking"—five categories, that are nifty heuristics for the scientist and equally, the artist/critic/designer, in *Ways of Worldmaking* (Hassocks, Sussex: The Harvester Press, 1978), 7–17. By name—*composition and decomposition, weighting [emphasis], ordering [derivation], deletion and supplementation,* and *deformation.* With scant that's definite enough to test, these may seem hopelessly idle; but my schemas help, with straightforward examples for each. Keeping to the sequence in Goodman's list, the addition schema $x \to x + t(x)$ and the identities $x \to x$ allow for composition and decomposition in terms of constructive methods and analytical ones, as in the second parentheses in A4, and in my early essay, "Two Exercises in Formal Composition," in note 44. Following this, the coloring book schema $x \to x + b^{-1}(x)$ at the start of A4 offers a useful slant on weighting, with shades and tones; graphs defined in retrospect are right for ordering, and likewise, so are hierarchies/trees, topologies, and kindred structures for derivations/calculations in shape grammars; the part schema $x \to prt(x)$ and its inverse $x \to prt^{-1}(x)$ work perfectly for deletion and supplementation; and finally, the transformation schema $x \to t(x)$ is the nub of deformation. Plenty connects these schemas, and many others are like them. For example, deletion and supplementation form a lattice. In a composition, $x \to prt(x)$ and $x \to prt^{-1}(x)$ define the schema $x \to prt(prt^{-1}(x))$ at the top, so that any change is possible; for any rule $A \to B$, it's simply $x = A$, $prt^{-1}(A) = A + B$, and $prt(A + B) = B$. What happens if this goes in reverse; what's in the schema $x \to prt^{-1}(prt(x))$? Try it for $prt(A) = A \cdot B$. The schemas $x \to prt(x)$ and $x \to prt^{-1}(x)$ highlight how rules take away from and add to shapes as they're tried, either schema after the other—whether I use a pencil or an eraser first or second doesn't matter. I guess this sounds intriguing enough for a quick example. Suppose the rule $A \to B$ is

in $x \to t(x)$, typically a translation or a reflection, but also a rotation. As drawn, the right edge of the square A in the left-hand side is the left edge of the square B in the right-hand side. The rule applies to A, so I can use my pencil to draw B next to it

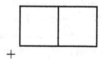

and then erase its horizontals and left vertical to get B. Or I can first erase the horizontals that are in A and its left vertical to get the vertical it shares with B

|

\+

and then use my pencil to draw B or just its horizontals and right vertical. The result is exactly the same drawing and erasing, or erasing and drawing. But there's still the lattice—x → x is at the bottom, to keep anything from changing, as identities pick out parts freely. This is easy to see in a diamond-shaped graph—

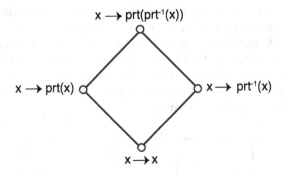

And there's more, if I add the erasing schema x → , to take away whatever I want to, and its inverse → x, to put in any shape anywhere, with the empty schema → at the bottom—a single arrow that's the see-nothing-do-nothing rule. Then, the bigger lattice goes in this way—

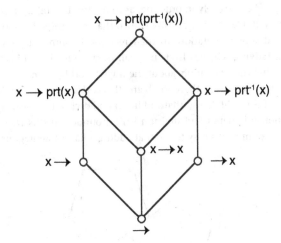

But sometimes, actual rules are a better way to see how things overlap in varied relationships. The rule

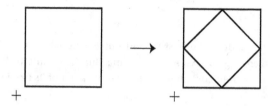

at the start of A1 is in x → x + t(x) and x → prt⁻¹(x)—is composition entirely supplementation? I guess it's better to try x → prt(prt⁻¹(x)) or x → prt⁻¹(prt(x)) to see what I get. It's tempting to say, decomposition is deletion—x → prt(x) includes x → x. In fact, identities may trace some parts and ignore everything else, as Alberti suggests for tree-trunks and clods of earth—and need I say, for drawings and pictures, etc. The part schema is simply x → Σ x', for each of the identities x' → x' used to pick out different parts x' of x. This leaves out or erases the part in the difference x − Σ x'. Moreover, the identity

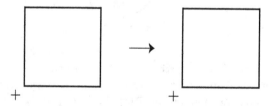

for squares is implicated extensively in ordering, as in A4 and Exhibit 2—it seems decomposition is really ordering, as well. I know, I've already shown that everything in Goodman's taxonomy is given in deletion and supplementation, in their reciprocal compositions x → prt(prt⁻¹(x)) and x → prt⁻¹(prt(x)). Classifying schemas isn't the goal; they're merely to illustrate Goodman's categories, heedless of what this may imply about the way(s) worlds are made. It seems that schemas resist final classification more often than not, being this and that, and something else at the same time. To get the feel of this, I like to conflate addition and division—at least sometimes. Addition is a Boolean operation for shapes, but division isn't grounded in this way, allowing for verbal sleights of hand. For example, it's easy to see that adding adjacent pentagons

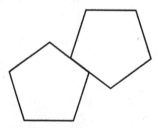

is the same as dividing their common area

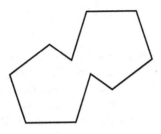

I can divide an L

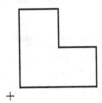

into three squares

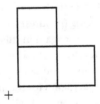

by adding two rectangles

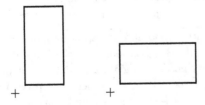

or even just a single square

The schema used to produce the Chinese ice-ray lattice designs mentioned in note 29 works in this way—adding polygons shows how sticks divide areas. Whether any of this is convincing or not doesn't matter. What's right is that schemas define rules that go together seamlessly, whatever I say they're for or think they do for the time being, coadunate in surprising ways whenever I calculate with shapes in the embed-fuse cycle. Calculating with shapes is the one category that may actually count—or is it two categories, when dimension $i = 0$ for symbols, and $i \neq 0$ for shapes?) There's room to spare to build lavishly on this with heuristics for pictures and poems, using schemas and rules for embedding and things that fuse, or maybe there are stronger methods and devices to conjure more impressive tricks, either in number and/or quality according to how I choose to parse—

to conjure more (impressive tricks) or to conjure (more impressive) tricks

Of course, this is never going to be stable; it flips back and forth as I divide and re-divide. I can never decide—it's exactly the same in figure-ground reversal, and in the unlimited multiplicity of the Rorschach test and in any beautiful form I like. There's just no denying it—language (meaning) is mostly a sublime mess in an ongoing process, markedly in trivial, everyday instances. But where would we be without ambiguity? At the very least, there would be no way to talk to anyone freely, including yourself—the precision (rigor) would be simply intolerable, especially if you're uncertain about what you mean, and are quick to change the subject as you go on. It's senseless to make up your mind once and for all, when it's so much fun to change it. That's the reason I prefer language (poems) to logic—and seeing (pictures) more. And isn't it funny how ambiguity is resolved as we continue to talk, over and over again in this way and that with no final result, but soon enough for mutual advantage, that brings in new insight and added understanding? I guess that's the reason for friendship, and conversation that breaks the rebarbative norms of mere communication—in commerce and in cybernetics for the animal and the machine. Conversation and communication define another ratio to use in place of imagination and fancy in my analogy

imagination : fancy :: shapes : symbols

I prefer talking to friends in conversation, raucous, noisy, and alive, and imagination and shapes in the embed-fuse cycle for shape grammars, to draw and paint—in conversation, ambiguity lets imagination soar, in communication it's refractory noise.

66. Dali urges seeing (delight) beyond all else, especially rationality and usefulness—physical performance and function/utility. He talks about the "paranoiac-critical method" time and again, in the way I like to talk about embedding and the embed-fuse cycle whenever I have a chance. In *The Secret Life of Salvador Dali*, trans. H. M. Chevalier (New York: Dial Press, 1961), Dali brings in "paranoiac metamorphosis," 304–305, and "paranoiac magic"—"negative hallucination . . . invisibility within a framework of real [physical] phenomena," 336–337. The first is ordinary embedding, while the second renders the embed-fuse cycle in terms of what doesn't happen, that is to say, when a part x isn't seen or vanishes, because a rule for it (an identity $x \rightarrow x$, or

generally, x → y) isn't tried. In shape grammars, shapes (parts) interact (combine) and disappear (add and fuse). I mentioned this early on, near the start of A1, when I described how rules work; moreover, it follows easily from the Law of Forgetting in the previous note. This is equally the Law of Invisibility—when I see this, I'm not seeing that. (It seems that all of us are pretty much blind in this way.) For a prescient gloss of embedding and shapes that fuse, there's also "The Rotting Donkey" in *Oui: The Paranoid-Critical Revolution: Writings 1927–1933*, ed. R. Descharnes and trans. Y. Shafir (Boston: Exact Change, 1998), 115–119. To Dali, "paranoid capacity," 116–117, is limitless, in keeping with the open-endedness of the embed-fuse cycle in shape grammars. (Paranoid capacity is Dali's measure of aesthetic value and grows with new perception, although this may seem incurably vague. The renowned mathematician George Birkhoff does better in the formula M = O/C—*Aesthetic Measure* (Cambridge, MA: Harvard University Press, 1933). The measure M of aesthetic value is the ratio of order O and complexity C, with values of O and C defined in terms of special visual analogies/spatial relations, and given transformations. And in fact, a fixed measure like this might work for Dali, as a panoptic summary of what there is to see, in a kind of average or integrated value. Maybe this involves a measure like C, for maximal elements—points, lines, planes, etc.—in the shapes that limit the parts that identities are able to pick out. As an experiment only, try the model shape

with eight maximal lines, their boundaries from x → b(x), and the maximal planes/areas they define/enclose in terms of the composition x → b⁻¹(prt(x)), for a value of 8 + 8 + 20 = 36—or some such averaged/weighted census. This isn't my tack in *Algorithmic Aesthetics*, where M is a model for $E_z$, note 29—in outlook, Dali and Birkhoff are poles apart, even as the one, in theory, welcomes the other. Shape grammars go for both, with Dali first and foremost.) This adds up to more than the "double images" that are taken for granted in figure-ground reversal, and regarded as amusing "illusions." It's ironic that for Garber (note 5 and note 46), they're enough—poetry and metaphor are inherently open-ended, but their creative sweep is conveniently circumscribed in the Rubin vase

for which paranoid capacity extends merely from the central vase to its flanking silhouettes. I was really lucky to find a copy (version) of the Rubin vase, mysteriously unnamed, in the chapter on paradox, in E. Kasner and J. Newman, *Mathematics and the Imagination* (New York: Simon and Schuster, 1940), 221, when I was nine or ten years old—I still know where it is on my bookshelf. I tried hard, but I couldn't see the opposing silhouettes. Was this a "negative hallucination"—the result of the Law of Invisibility? It took several days of concentrated effort, going back to the vase time after time, before the faces abruptly kicked in. And my work panned out; when I finally pulled it off, the trick changed the order of my universe. I remember how it felt to see new images switch back and forth—they were a truly marvelous discovery. This kind of seeing was totally amazing; it was something special that I could try everywhere to re-create (invent) my world in ongoing free play. With a little effort, it put everything in motion to swerve in any way I pleased. For example, the two A's side by side in A3—the ones with a raven and twin peaks—are like the Rubin vase; the V (beak) in the raven is the central vase, and the adjacent A's are the flanking silhouettes. This works when there are vertical cuts (axes) to pick out pieces (parts) in alternative ways. And it's a cinch to find more in the Rubin vase—try the boundary schema $x \rightarrow b(x)$ on the silhouettes, to revert to lines only in the shape

maybe a tripartite, decorative screen with two panels open in front, or is one panel of the screen unfolded in front and the other one in reverse in back, to make a zigzag? How about other parts— this one

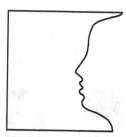

overlaps its reflection

Now where are the profiles of the faces, and how are they oriented? Maybe they're daydreaming/gawking/spying, as they look out/in an open/closed window (James does this for a line running east/west, when he discusses the "human element" in familiar objects, see *Shape*, 152, in sync with Coleridge's synthesis/indifference)—but try it again, lower down in reverse. I wonder how the combinatorics plays out. And how do things go if I use the coloring book schema $x \rightarrow x + b^{-1}(x)$ on the seven bounded areas in

so that shades of grey vary as these areas overlap in multiple ways? (Alternatively, try the schema $x \rightarrow x + b^{-1}(prt(x))$, for a single value of x.) There's plenty more like this in the Rubin vase, in A's, and throughout Exhibit 2. Every one of us is free to see farther, whenever we please and entirely as we please—even if the uncertainty is a relentless source of anxiety. Dali embraces the vast (sublime) chaos and confusion of shapes and their parts, never stopping to think; his eyes are wide open instead, vigilantly searching for more. The terrible risks in this are wonderfully clear in the campy visual wit of W. H. Auden, in the "Journal of an Airman," Part II of *The Orators: An English Study*, in *The English Auden: Poems, Essays, and Dramatic Writings 1927–1939*, ed. E. Mendelson (New York: Random House, 1977), 74. The passage is one of many things that has to be drawn and seen. As Auden shows, sometimes words (visual analogies) alone won't do; they leave out much too much—there's more to see than can be said. Sooner or later, the eye (perception) is indispensable and best—

Give the party you suspect the figure

and ask him to pick out a form from it.
If he picks out either of the two crosses

you may accept him as a friend, but if he chooses such a form as

it is wiser to shoot at once.

There seems to be a rule for this—a good form is a familiar one with an easy name. Otherwise, how do you know what it is—it can only be shown, as you duck for cover in a fox/rabbit-hole. To be safe, seeing must be verbal—see only what you can say (categorize). It's one way for the mind to suppress the eye. I guess something like this also goes for poets (artists and other criminal parties) who try revisionary ratios (schemas and rules) on their precursors, and at times, on themselves—there's an overwhelming need to break free, in crossing lines of fire. This isn't too unlike Bloom's notorious anxiety of influence, in the note just before this one—when new poems breach the bounds of prior work, when imagination takes hold in a vibrant, more expansive present that assimilates and reorders the past in surprising ways. Auden's drawings are a nifty example of embedding (poetic misprision) at work, in an everyday process that doesn't make a definite law out of it. This is exactly how it should be whenever there's anything to see—once bad habits and rote training are overcome, chances for fusing and new embedding are pretty good most of the time; they're not to be avoided. And how many more different (creative) ways are there for Auden's airman to shoot—to take a bullet tracing the varied parts in his figure, and elsewhere in all manner of places? The odds for a fatal mistake are daunting, even when things are as cut-and-dried as Auden's forms seem to imply, so that only unit segments defined by intersecting lines add up, to combine in visual analogies. (This amounts to a canonical design space; it's neat how it coincides with standard practice when computers are used in design, conspicuously in engineering. The unit segments are 36 in all—six for each horizontal and vertical, and three for each diagonal, but only 20 shortest segments to partition forms into $2^{20}$ distinct classes, including symmetrical copies.) If I were there, I'd play it smart and cover my eyes or better yet, I'd pretend to be blind. Why place a sucker's bet on such a long shot? Come on, take a look and have a go—the airman might miss. What happens if I pick out the ten shortest segments in the figure that form a black widow, a decorated war veteran that's lost four of its eight legs

—or maybe its missing legs are concealed (embedded) in the hourglass or somewhere else. And suppose I keep the other ten segments for the spider's hapless better half

Then again, I can try something more resourceful—how many other ways can I pick out the spider and its spouse? With the unit segments from Auden's figure, the spider is as few as six segments—how many legs are there now?—and as many as 14, with multiple arrangements for every number in between. What do you see just by looking at the spider, in the following trio of arrangements? How does the spider change as its segments vary in number from six to eight or ten, in this arrangement or that one—

That's the trouble with visual analogies and underlying (hidden) structure—they're independent of seeing. (This is also a good reason to partition design spaces.) Surely, the perversity (depravity) of forms is evident—but for which ones and for how long? What's strange now needn't be that strange the next time I look, with a list of brand-new names ready to try. And what happens when segments aren't units, and new units aren't defined—when everything is a Rorschach test or a beautiful form? What happens if I take a second turn after I've used the part schema $x \rightarrow prt(x)$ to pick out a form? What if I use its inverse $x \rightarrow prt^{-1}(x)$ to put in more—seeing is adding, too. (In Goodman's taxonomy of worldmaking in note 65, x → prt(x) overlaps deletion, and x → prt⁻¹(x), both composition and supplementation.) How hard is it to alter the form? Is there a special knack to it? Do identities x → x do the trick? Does adding to the form repay the effort or not? Once changed, is the form charged with meaning? Is the spider's spouse in a new form/guise, the drawing of a friend or foe? (Who did the drawing; who/what does it show; how can you ever know?) Maybe a cunning secret agent subverts the rule of good form with the schema x → prt⁻¹(x)—IT'S A BOMB

Auden's test seems magically prescient, with a difficult (ambiguous) measure of success that aligns perfectly with Turing's famous test nearly two decades later in 1950, to decide if machines can think, and the picture tests used on some websites today, to decide whether you're a real person or another annoying computer bot. No doubt, Auden's airman will manage to make good sense of the drawings in the following series of three, in L. Wittgenstein, *Remarks on the Foundations of Mathematics* (Cambridge, MA: MIT Press, 1967), 189e. It seems that the airman and a famous philosopher are worried about similar things—what difference does it make when what I see changes? But the airman may have a better chance to figure this out in a reasonable (coherent) way. And if he can't—well, his weapon is charged, and he's ready to fire at will. Is this how philosophy ends, after uninterrupted millenniums—another unsuspecting victim of a few stray lines? Count them. Is the addition for symbols (sets) or shapes—is the answer to six plus three, six or eight or ten, or is it some other number? My teachers told me it was nine. Who says you can't argue aesthetics—

An addition of shapes together, so that some of the edges fuse, plays a very small part in our life.—As when

and

yield the figure

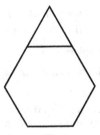

But if this were an *important* [life and death] operation, our ordinary concept of arithmetical addition would perhaps be different.

Now you might of course say: "In this case the manipulation of figures according to rules is not calculation."

But then, what is it? Does calculating exceed visual analogies in the Rorschach test and a beautiful form? Imagine the entry in the airman's journal to separate (classify) friend and foe, in a life and death decision for everyone involved. Isn't the A obvious, when the hexagon and the little triangle on top of it fuse and then re-divide? The symbol/shape I started out with in my ABC of seeing, after Minsky and McCarthy and Chomsky in A3, pops out in a surprising way, and will again in Exhibit 2—to show the inevitable failure (incompleteness) of visual analogies. And how far does this actually go; is there no end ever in sight? Isn't there a spider in the figure, or one-quarter of one, in five lines

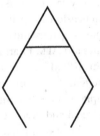

Not everyone is ready for the constant risk of trying rules in varied ways in the embed-fuse cycle. But for sure, Dali is—in fact, paranoiac (a madman), or is this merely the sign of genius? Any trace of the one (paranoia) or the other (genius) is sufficient reason to stand up, take notice, and act. (Both Wilde and Turing can attest to this—it's not always fun being a genius, and may be life-threatening.) In uncertain times, it's wise to shoot at once, or to run for your life. There are too many surprises and dire consequences to take any unnecessary chances—isn't it better to be safe than sorry? Genius in art and design is a wonderful goal, but it isn't worth the gamble when the odds are stacked against it.

67. Wilde, *The Uncensored Picture of Dorian Gray* (Cambridge, MA: The Belknap Press of Harvard University Press, 2011), 78.

68. In *The Making of a Philosopher: My Journey through Twentieth-Century Philosophy* (New York: Perennial, 2003), 206, Colin McGinn describes calculating in an awkward formula

> Combinatorial Atomism with Lawlike Mappings

with the cheesy acronym "CALM." Still, the handful of pages on how this works (pp. 204–212) is impressive; a few words set the scene—

> Nature is a system of derived entities, the basic going to construct the less basic; and understanding nature is figuring out how the derivation goes. The CALM structure is the general format for this kind of understanding: Find the atoms and the laws of combination and evolution, and then derive the myriad of complex objects you find in nature. If incomprehension is a state of anxiety or chaos, then CALM is what brings calm, 207.

Once atoms are found—don't ask how (see note 76)—Lego (compositionality) of this kind traces the locus of understanding; in McGinn's CALM conjecture, it separates what's intelligible from what's not, in terms of cognitive (hierarchical) structure and our biological makeup. Maybe, but not for the artist/critic—Simon risks "combinatoric play" and visual analogies to take more from painting than CALM allows, and Garber puts metaphor past atoms and mappings in ambiguous figures. Painting and metaphor unfold in vast confusion, in von Neumann's pictures and Rorschach test, and Wilde's beautiful form. Nor does CALM suit the philosopher—"The essence of a philosophical problem is the unexplained leap . . . not dictated by what precedes it . . . we seem to be presented with something radically novel, issuing from nowhere, as if a new act of creation were necessary to bring it into being," 209. (Cf Chomsky on creativity in language in note 47.) Yes, art and design go like this—and farther. For the artist/critic, imagination (to re-create) is calculating with rules in the embed-fuse cycle, and working things out for the time being, step by step in reverse, in tentative comprehension to spell MLAC in retrospect. Bergson tries this sequence for scientific discovery in note 7—genius and invention succeed before knowing how. Whether it's better to follow the dictates of CALM in art and design, calculating/understanding with atoms and mappings/recursion in AI/BIM, etc., or to adopt new ways in the embed-fuse cycle—that's the question. On the one hand, McGinn's version of nature is fancy (physics, for example, is a "discrete infinity" of complex entities that interact in space, tailored for schools and STEM); nature's objects as objects are "fixed and dead," becalmed in the doldrums of cognitive structure, and "'sicklied o'er with the pale cast of thought,'" 213 (to quote McGinn recursively, quoting Hamlet in *Hamlet*). And on the other hand, calculating in shape grammars "puzzles the will" in open-ended ambiguity and indeterminacy; the vagaries of the eye "make cowards of us all"—adrift in reverie, "what dreams may come" as shapes unfold? Endless in the embed-fuse cycle, perception (seeing) ignores the lingering musts of prior structure to best thought (atoms and recursion). (The examples in Exhibit 2—in particular, in note 17—show how shapes elude CALM, and subsume it with ease when rules are tried in the embed-fuse cycle.)

69. My book *Shape* is a current account of shape grammars, and how they're used for visual calculating. It's illustrated lavishly (Mine Ozkar turned my rough sketches into finished drawings—more than 1,100 of them by one count) along with elaborate formal detail on shape algebras

and like mathematics, and extensive background and history with roots in philosophy (James's pragmatism) instead of art and literary criticism (Coleridge and John Keats, and Wilde's critical spirit). The real truth is that I'm far more relaxed with the ambiguity in pictures and poems, and its origin in insight and imagination, than I ever have been with the anxious clarity that's *de rigueur* in philosophy. I've always found pure delight in negative capability—and being messy. But it's just a feeling—Knuth maps another way in computer science (see note 22) that some philosophers are inclined to take. Rapt in thought, they're sure that any theory of imagination (creativity) is a pointless undertaking, no better than a theory of Tuesday—yes, there's a theory in each case, but merely a tiny one lacking in seriousness and scope, and unlikely to hold much of value. (That imagination and Tuesday are the same in theory is from the philosopher and cognitive scientist, Jerry Fodor. C. K. Chesterton, however, finds more in this relationship for all the days of the week, in an ever-expanding mystery, in *The Man Who Was Thursday: A Nightmare* (New York: Dodd, Mead & Company, 1935). My mother gave me her copy to read, around the time I was trying to figure out the Rubin vase, in an uncanny coincidence. Never mind seeing and ambiguity—I think she wanted me to notice something more. Definite things aren't definite forever; invariably, they're different now than they were before. Meaning relies on change in an elusive flux, and is empty without renewed surprise. Growing up with this in mind was my start on negative capability—of course, at home and not in school, where learning was rote. It was a piece of cake in Hollywood. Once I got the true feel of it, there were surprises almost everywhere I looked; my friends told me to ignore them, to keep calm, and to settle for what I knew, but I decided not to take their advice. Surprises made things pulse, and were too much fun—ambiguity and meaning were right for insight and imagination.) Negative capability is probably why little of what I said in *Shape* is explicitly in my answers to questions Q1 to Q7, although it's worth comparing *Shape* and my seven answers for overlaps—and they're extensive and many. Some of the things I say in A1 to A7 may seem to be contradictory. I was careful about this in *Shape*, but only a few took heed. My impulsiveness now is a kind of expository experiment—an inconstant difference that may help to make visual calculating in shape grammars more accessible to a wider audience. I tend to think of every sensible question as a Rorschach test, so everyone can participate at any time in his/her own way. All of us are off the hook, to answer/see as we please—and this goes especially for me. I've made no commitment to keep to anything I've said before, particularly if it's just for foolish consistency or tedious coherence. Consistency and coherence are only good for logic and definite (yes/no) answers; they go for scant at the quick of art and design. This strikes me as entirely in keeping with the true spirit of shape grammars—to see things as in themselves they really are not. And this is impossible without imagination and the embed-fuse cycle. It would be remiss of me not to remark that *Shape* is an effective answer to *Algorithmic Aesthetics*—seeing and other modes of sense experience are correlative to schemas and the rules they define for use in the embed-fuse cycle. There's no reason for anything to be out of sight or hidden from view—that's the central claim in shape grammars that's somehow lost in algorithmic aesthetics. In *Shape*, algorithms (calculating) and aesthetics hinge on new perception (insight and imagination)—everything is von Neuman's Rorschach test, and equally, Wilde's beautiful form. *Shape* is for seeing and rereading in myriad ways—G. Stiny, *Shape: Talking about Seeing and Doing* (Cambridge, MA: MIT Press, 2006).

70. O. Wilde, "The Decay of Lying," *Intentions*, 299. It's easy to feel lost and rather at sea reading this or the lines on p. 41 (see note 48 for details), or trying to follow Wilde's aesthetic formula, when aesthetic experience is open-ended and entirely your own. For the utterly bewildered, there are safe options couched in reason. Hume charts a true course through personal misdirection and varied points of view, for a practical standard of taste and beauty—in D. Hume, "Of the Standard of Taste," in *English Essays from Sir Philip Sidney to Macaulay*, ed, C. W. Eliot (New York: P. F. Collier & Son, 1910), 215–234. This orders two extremes in a refractory distinction, so that the first tempers the second to keep it from spinning completely out of control—

> The difference, it is said, is very wide between [1] judgment and [2] sentiment. All sentiment is right; because sentiment has a reference to nothing beyond itself, and is always real, wherever a man is conscious of it. But all determinations of the understanding are not right; because they have a reference to something beyond themselves, to wit, real matter of fact; and are not always conformable to that standard, 217.

No one doubts that judgment is stern, tied firmly to independent facts and prior data, and to reason and use—facts and data in understanding are the sine qua non of AI and machine learning. But how real (steady) are facts and data in poetic activity—in shape grammars and visual calculating in the embed-fuse cycle? Can observation (seeing and looking) alter facts? Are there triangles when I draw squares, or squares in triangles and in other distinct polygons and letters? Is the hapless better half of a black widow snared in a bomb? Such tricks abound in visual calculating—"How easy is a bush supposed a bear!" Parts are evanescent in shapes of all kinds; they're divided anew every time a rule is tried, comprehensively only in retrospect, whenever calculating ends, at least for the time being. Things that change like this—at will, from what they are to what they aren't—evoke sentiment and arouse aesthetic imagination in ambiguity, divergent points of view, and enduring strangeness. The immeasurable surprise aligns with taste and beauty, nicely in Hume's sense—

> . . . a thousand different sentiments, excited by the same object, are all right: Because no sentiment represents what is really in the object. . . . Beauty is no quality in things themselves: It exists merely in the mind which contemplates them; and each mind perceives a different beauty. . . . According to the disposition of the organs, the same object may be both bitter and sweet; and the proverb has justly determined it to be fruitless to dispute concerning tastes. It is very natural, and even quite necessary, to extend this axiom to mental, as well as bodily taste, 218.

And Wilde adds to this—"Beauty has as many meanings as a man has moods." There's a swerve from judgment lodged in facts (abstractions in descriptions/representations that fix what's really there, maybe in parametric variation) to individual sentiment (mental taste), with scant in judgment steady enough to "modify and restrain," 218, Hume's perceiving mind, and the course of Wilde's critic/artist adrift in reverie and prone to fickle moods. The arc goes the other way around—taste and beauty precede facts; sentiment (aesthetics/perception) informs judgment (reason/use) as shapes (dimension i ≥ 0) inform symbols (dimension i = 0). The derivative analogy

sentiment : judgment :: imagination : fancy

follows to suggest another relationship for sentiment and judgment in facts and data, and to affirm the centrality of taste and beauty. But Hume has something less conjectural in mind than the embed-fuse cycle in shape grammars, elaborated in analogies—"our intention in this essay is to mingle some light of the understanding with the feelings of sentiment," 222. In a quirky way, this is combinatorial in keeping with Hume's constructive account of how complex ideas are made up of simple ones—truth, or so it seems, is forged in a consensus of separate critics. (The same kind of thinking is easy to find in AI; for example, it's a small step to Minsky's "agents" or "particles" of thought that combine and interact in groups and hierarchies in "the society of mind"—conflict and competition keep agents from working together, while compromise and cooperation encourage consensus in complicated and largely unknown ways. In contrast to this, rules in the embed-fuse cycle neither clash nor require reconciliation; they apply in concert without negotiation and accommodation, either one at a time or many, indistinguishably—there are no facts in descriptions and representations, in hidden/underlying structure, to modify and restrain how rules are tried, only the feelings of sentiment remain in ongoing perception. Maybe visual calculating is more original than I previously thought; it's hard to find anything like it in conventional computer science—I guess that's it for AI/BIM and parametric variation.) Hume sets a pretty high bar for critics; these demigods and guardians are as refined as the decisions they make—

> Strong sense, united to delicate sentiment, improved by practice, perfected by comparison, and cleared of all prejudice can alone entitle critics . . . and the joint verdict of such [rare few], wherever they are to be found, is the true standard of taste and beauty, 228.

Today, critics are philosophers and scholars, and others of strong sense, etc.—aesthetes of all sorts, including art historians and pundits, museum curators and docents, and maybe auctioneers, wealthy collectors and connoisseurs, and trendy gallery owners. Their verdicts as jurors hang on museum walls and are listed in catalogues; they're on the web and in the classroom, and discussed at openings. This goes best with a little supervision and oversight—what juries decide can be checked for uniformity by polling the critics singly, and for influence, in random surveys, and on multiple-choice tests for students. There's the ring of truth in brazen circularity—entitled critics show the rest of us how to see and what to feel, and use our collective support in loyal yeas to reaffirm and verify their verdicts, and their special remit as critics/jurors. Somehow, everyone gets a say in the reciprocity of verdict (decision) and rote response—not all sentiment is right. Computer scientists love this; they rush in rubbing their hands in greedy anticipation, because there are endless facts and data in polls, surveys, and tests, to probe and measure art and design in a convergent standard of taste and beauty. And before you know it, computers take over completely, to replace the critics in automatic decision making—AI and machine learning make sure of it. Nonetheless, it's a mystery that a standard of taste converges to a verdict. I guess that's why this standard isn't safe with all of us—it isn't much of a standard with checkered results. Not everyone is qualified to be a juror (expert). Whatever critics decide, I prefer to look on my own; sooner or later, I need to see for myself. I'm keen on individual sentiment and Wilde's aesthetic formula that defies standards, and von Neumann's parallel riff on pictures and the Rorschach test in the sway of personality—on calculating without facts and prior data, when calculating alters to include sentiment (perception), so that things aren't as in themselves they really are in visual analogies and spatial relations. The ultimate slur of uselessness goes hand in hand with this, to bear in unbroken silence with no respite or redress—but lacking the chance for a true standard of taste and beauty isn't hopeless. There are rules to try, for example, identities $x \rightarrow x$ to change things by looking, and other rules in the schema for parts $x \rightarrow prt(x)$ and its expansive inverse $x \rightarrow prt^{-1}(x)$ to take away and fill in, to please yourself in different ways whenever you like. Everyone has rules that anyone can use at any time to show us "the whole fiery-coloured world" as it unfolds in the fluctuating give and take of how to see and what to feel, and to argue aesthetics in wonder and delight. This isn't rudderless—rules in the embed-fuse cycle make sure of it, for everyone to see in a special way, from any point of view. The numberless feelings of sentiment overflow in new perception, adrift in reverie, caught in currents of inconstant moods, and immersed in imagination.

71. O. Wilde, "The Soul of Man under Socialism," in *The Artist as Critic: Critical Writings of Oscar Wilde*, 255–289. Wilde assures us in his wonderful way that personal wealth is an unwelcome burden. In order to keep wealth, you have no choice but to take care of it, and to nurture it—

> Property not merely has duties, but has so many duties that its possession to any large extent is a bore. It involves endless claims upon one, endless attention to business, endless bother. If property had simply pleasures, we could stand it; but its duties make it unbearable. In the interests of the rich, we must get rid of it, 258.

Of all the arguments for Socialism I know, this is the strongest—and the support Socialism gives to art and design for pictures and poems. It's simply amazing how much effort it actually takes, how much energy is senselessly lost, to protect capital and property, and to multiply numbing wealth. And as stores of money continue to compound in one Libertarian scheme after another, the task does, in fact, become "endless." It's certain that imagination (esemplastic power) isn't for the rich—they have no time for it, as they count and recount increasing hoards of lucre and gold, and hasten to deplete their souls. ("Getting and spending, [they] lay waste [their] powers.") Wilde casts economics in a new light that frees it from business and bother, and the muck and mire of dismal data. Socialism is his way around numbers (units) and counting to spare everyone the pain and toil of wealth and poverty alike, to enrich all souls—imagination supersedes fancy with its given counters, and fixities and definites, as it re-creates strange wonders in pictures and poems, and in beautiful form.

72. Ambiguity is noise. At least they're easy to confuse—"When the same person judges [looks at] the same object many times and reaches [sees] different conclusions [things], that's one kind of noise [ambiguity]." It seems that the vagaries of appetite, mood, taste, and whim riddle personal judgment/perception—"Noise [ambiguity] in general is unwanted [surprising] variability." And this goes for any "system" of multiple individuals—maybe in courts when judges hand down different opinions for identical facts. "We look at the same world [try the shape

for example], and we look at it with confidence. I feel that I'm right in most of my judgments, and I'm truthful to you. I respect my colleagues, and I like them. And they are looking at the same world. I expect them to see the same world that I see. But in fact, they don't. That's the surprise." There's endless delight when seeing doesn't matter, but doubt and insecurity spread when it makes a difference in our lives. A proven method that works for every problem provides a sure way out; the effective suppression if not the total elimination of ambiguity and noise to guarantee fairness (consistency/uniformity) and such is practicable only in AI and machine learning, in D. Kahneman, O. Sibony, and C. R. Sunstein, *Noise: A Flaw [sic] in Human Judgment* (New York: Little, Brown Spark, 2021). Computers deserve our lasting trust—"this is true in general of artificial intelligence, their capacity for learning is incredibly higher than humans, which is why when an AI, an artificial intelligence come[s] close to human performance, you can bet with high confidence that within a few years, the AI will be better." And this is good news for Socialism; the day isn't too far off when AI will do what we do better than we can—it's so long to appetite, mood, taste, and whim. Of course, AI extends only to useful things; it goes without saying, which is why I'm saying it, that AI won't do for calculating in art and design, where ambiguity—seeing things as in themselves they really are not—is the very heart and soul of insight and imagination in the embed-fuse cycle. Computers take away the looking—and with it, creativity. Is there any delight—or fairness for that matter—when no one can see? What kind of world would this be without noise, with no chance for change or to change your mind at will, with neither hope and joy nor regret? There's plenty of room for AI to do the useful (rote) things we do—and much, much more for us to see in fickle ways and thrive. Ambiguity and noise aren't a flaw in judgment for computers to correct, but an opportunity for calculating to seize. (The quotes are from Daniel Kahneman on *Noise*, when Kara Swisher interviewed him for *The New York Times*, May 17, 2021.)

73. It's truly remarkable how great philosophy so often misaligns with art and design. More than two decades ago when I was at UCLA, I had the good luck to show a famous logician/philosopher how visual calculating worked in shape grammars—it was a perfectly delightful conversation, full of generosity, and give and take. He was really amazed that the shape

looked different as two squares and four triangles, and couldn't stop talking about how neat it felt to switch back and forth at will—the discontinuity in number and kind was sheer magic, and the nub of seeing. Nonetheless, he made sure to point out that the embed-fuse cycle was suspect and most likely a fraud. Wasn't I the one who described identities x → x and other rules as "tricks," with no basis in reliable (eternal) principles and laws? It never occurred to me that this might be a problem. The embed-fuse cycle with reduction rules, etc. seemed perfectly rigorous to me, at least as rigorous as Turing machines and logic, and easier to grasp by looking. I said not to worry—"In *A Midsummer Night's Dream*, Shakespeare uses Theseus to talk about 'strong imagination' in terms of tricks"—

Such tricks hath strong imagination,
That if it would but apprehend some joy,
It comprehends some bringer of that joy;
Or in the night, imagining some fear,
How easy is a bush supposed a bear!

The tricks of strong imagination, it seems, are the root source of cause (knowledge)—the "bringer of that joy"—and the excessive effects of the eye (perception). And to my surprise and relief, this outburst of romantic certainty in a poetic reference—in fact, blatant name-dropping—did the trick for dthat[the philosopher I see standing to my left]. There was scant need for Coleridge's esemplastic power; it seemed right to put it aside for the time being, for something more than slippery semantics, and shifting scruples and qualms. Plato is a different story; he brooks no sway—you're with him or not in a binary choice, yes/no, yea/nay, true/false, 0/1, etc. On the one hand, objects (pictures and poems) are the same forever, with nothing to see that hasn't been described or represented before, for example, in visual analogies that substitute for pictures and thereby preclude open-ended perception; and on the other hand, objects are vital, transient in motion and flux, and entirely indifferent to the way they're described here and now. The passage I like to return to, over and over again, is from, *The Republic*, trans. B. Jowett (New York: Vintage Books, 1991), 214—

> Inasmuch as philosophers only are able to grasp the eternal and unchangeable, and those who wander in the region of the many and variable are not philosophers, I must ask you which of the two classes should be the rulers of our State?

It seems that the invariant orders perception in terms of positive norms. Isn't this the same for flawless judgment in note 72? The evanescent and changeable, however, are easy to miss, and if noticed by accident or chance, they're as easy to ignore and forget. To revel in ambiguity and the unexpected, to alter what you see freely, as you see it, is a kind of negative capability. I thought to myself—what a marvelous way to start *Shape*, 1, inasmuch as seeing is the primary locus for imagination in art and design. But many who read my book or parts of it were totally baffled when I balked and didn't answer unambiguously, "philosophers," in the way Plato's passage justly implies—to affirm the eternal and the timeless in a binding ritual of final participation that's entirely fancy. After all, wasn't I calculating? The answer was obvious and the quote was trivial, just not for visual calculating and how it works in shape grammars. Maybe experts in computers and AI (symbolic calculating) want to embrace the eternal—there are a lot of people around MIT and elsewhere, in philosophy and in art and design, too, who toy with the idea of eternal things, even themselves. No matter, the temptation to wander is ineluctable and impossible to resist, when there's so much more to see that's pure joy. Who, in his/her right mind, would choose of his/her own free will to give up his/her soul—to forswear imagination and delight in ambiguity, only for the eternal and eternity? Barfield's final participation, even as it revivifies prior perception, isn't what it's cracked up to be—anything that's invariant for too long grows stale, and without change, it loses its meaning. It seems that in the fixity of eternal things (maybe Dorian Gray), there's only instability, at least when you follow your nose—the decay can't be ignored. Nothing is timeless at the quick of art and design (but cf note17). Alberti caught on to this many centuries ago—for the winged eye, there's more to see whenever and wherever it swerves, always in motion in unencumbered flight, wing beat after wing beat in the embed-fuse cycle.

74. O. Wilde, "The Critic as Artist," *Intentions*, 365.

75. Ankur Moitra, a computer scientist at MIT, puts it in this way—"Machine learning is eating up the world around us, and it works so well that it is easy to forget that we don't know why it works." M. Jarvis, "On a Quest through Uncharted Territory," *MIT News* (June 6, 2021). Who needs half-knowledge (negative capability) when no knowledge (machine learning) will do? I guess that's it for ambiguity in the embed-fuse cycle—so long to imagination.

76. Near the beginning of his conversation with Ulam, Rota, 56–57, repeats a similar kind of infinite regress—Descartes's famous homunculus problem at the root of consciousness, that's unsolved in AI, cognitive science, and elsewhere. (The same goes for Russian dolls and infinite loops, and maybe in note 68—in McGinn's CALM, "understanding" how atoms are derived from atoms, derived from atoms, derived from atoms, etc. This makes it hard to keep calm about CALM, and underlying structure in general; sooner or later, Lord Henry is right—judgment is superficial, lodged solely in appearances. Still, finding atoms/structure may be part of our special gift for theories, from Shakespeare's tricks to Peirce's abduction, in an instinct/knack for explanation that's mostly a mystery. But not for shapes in the embed-fuse cycle—then atoms and structure are resolved in retrospect, redefined every time a rule applies.) AI ignores explanation and often consciousness for common use, lucky when things work no matter why. I rely on engineering for how and physics to understand how and why—with more to say that saves the appearances, in the nano-interval from lame excuses to coherent theory.

77. H. Bloom, *Shakespeare: The Invention of the Human* (New York: Riverhead Books, 1999), 718. The leading epigraph for *Shakespeare* consists of a few words from Friedrich Nietzsche's *The Twilight of the Idols*, in Bloom's personal translation; it casts a jaundiced eye on Bloom as critic and on the Bard as artist, and questions the role of language for all of us—

> That for which we find words is something already dead in our hearts. There is always a kind of contempt in the act of speaking.

It's profoundly dispiriting to discover that words limit the things (pictures and poems) they describe, sapped of interest and meaning once they're named, so that they're nothing but . . . , suited for naught besides definition. But Nietzsche seems perfectly right for Coleridge's fancy— that objects as objects are fixed and dead, permanently snared in counters or fixities and definites. What we say about things in descriptions and representations in order to use them as objects is comically/tragically incomplete, and contemptuous because in our hearts, we know that it betrays appearances and impressions (sentiments) to replace them with something already known—everything important remains unsaid, to unfold only in an ongoing process in which things combine and alter. (The third example in Exhibit 2 provides another take on this in terms of Pythagorean triangles and squares, and A's and squares.) Why do we give in so abjectly—for the sake of symbolic calculating in graphs, and mappings for metaphor, for the convenience, expediency, and security of visual analogies and spatial relations in tamper-proof sets, for the promise of inclusive (universal) participation in shared structures (idols or collective representations, taxonomies, Hume's standard of taste and beauty, etc.) that must be taught to everyone and then enforced, for the far-fetched claims and fantastic dreams of the latest technology in AI/BIM? Will fancy do—to see things as in themselves they really are not? Does its locus in "combinatoric play" with building blocks (counters), whether richly varied or not, bring in all of art and design? Can this ever complete the soul? Maybe the scientist and the computer expert don't care, dashing in their splendid hats, satisfied with mere data and learning alone, in a compulsive quest to tame the world with more and more. But the eye "flash[es] in every direction," as it swerves in a wanton course, independent of data and learning; imagination and esemplastic power change things to charge them with meaning—to see past/through what's been said before in descriptions and representations. In shape grammars, the embed-fuse cycle pulses as rules apply to see and do in an ongoing (recursive) process that brooks no bounds, to observe, and add and take away in shapes and words (symbols) alike, to re-create at will, so that things are "essentially *vital*," radiant in constant motion—perception and such tricks raise the dead, in the extravagant magic of the embed-fuse cycle. Hallelujah—isn't this proof positive? What more can I possibly say, to get you to take visual calculating in shape grammars seriously—to open your eyes and see? For God's/Nietzsche's sake, there's "the whole fiery-coloured world." Nietzsche is rendered better, in another way in shape grammars—

That which the eye finds anew is something at once alive in our hearts. There is always a kind of joy/love in the act of seeing.

Even so, visual calculating seems way too optimistic (frivolous/naïve/trivial) to be of any value, given the rampant social injustice and environmental pessimism of the day—especially when computers with remarkable speed and vast memory lay claim to a better world. A one-way swerve (retreat) from shapes to symbols is required for real (scientific) progress; it takes away the looking in confident analysis to facilitate computers, discounting aesthetics in the name of practicality, reason, and whatever is just and fair. (How analysis works in the swerve from shapes to symbols is pretty much a mystery, outside of shape grammars. The way symbols evolve as rules interact in shapes is a retrospective byproduct of calculating, yet this is typically dismissed as mystical, something left for art and design, and mocked in science—maybe because thinking only begins after symbols are defined. It's undeniable that words and sentences are how we make up thoughts in our minds, and how we convey our thoughts to others whenever we please—but what about pictures and poems?) Once again, delight in appearances is diminished in the Vitruvian canon—in Ruskin's decoration/ornament that exceeds ordinary function and use. Unlike the fickle eye, the rigorous mind can't be gamed—in the architectural scientist's encoded built form that makes counting easy. But lacking mutual back and forth for shapes and symbols, counters merely harden our hearts, combining predictably in terms of explicit axioms or ones implicit in stores of prior data to create new things locked rigidly in "verbal" perception—a shape is its name, no more and no less, encoded fully in visual analogies, spatial relations, etc. Of course, this isn't as straightforward as it sometimes seems; it's what von Neumann worries about in pictures and the Rorschach test—that there's more to a shape (a triangle) than symbols (three edges and a name) allow. The computer scientist has plenty of unfinished work to do, to reconcile seeing and saying (encoding). It's an elusive if not impossible goal—an easy example begins to show why. Two triangles, one two (four) times the size of the other

and three triangles, all the same size

combine to make five triangles, one large and four small

The figures in this trio are visually identical, and numerically distinct. I can delete two triangles from the first, and not three; and I can delete three triangles from the second, and not two that differ in size. However when I delete two triangles or three from the third, nothing changes that I can see, so I can't decide what I've done without trying to delete more, in a forensic test that requires logic and memory. (Friends in AI tell me that inadequate symbolic memory is a key deficiency/flaw of the human mind—so much the better for the artist and poet, who revel aimlessly in open-ended perception.) Logic and memory take away the looking, with no comprehensive way to restore it in words anywhere in sight. These difficulties aside, and there are plenty more that widen the gap between seeing and saying, this is the universal way to calculate in computers—those in the know are sold on counters in words, etc., as a must for recursion. (To me, 0's and 1's—the prerequisite "atoms of computation"—are Turing's curse, although I'm pretty sure Turing would find shape grammars perfectly fine, along with the difference that dimension makes in calculating, for $i = 0$ and $i \neq 0$.) And when computers are up and running, they're a tremendous boon. Nonetheless, the resulting technology is an empty lure, smug and bereft of surprise—two triangles in words aren't three or anything else, no matter how they look. (The initial two-thirds of Exhibit 2 expand on this, with more on triangles that combine in twos and threes, and how they're related, and for rows of squares in overlapping rectangles.) There's neither art nor design in Nietzsche's words (symbols) for things—art and design are lost in a murky past—and no room for the individual artist or critic to see and to thrive. Words and symbols create (generate) many things and are used to describe them in the same way. Computers promise an efficient and ordered world that overlooks delight and its source in ambiguity, inherently tethered to counters in fixities and definites. This will not quite be a culture—or even approximate one—forlorn, without imagination and esemplastic power to tend the soul.

78. In *The Anxiety of Influence* (see note 65), Bloom calls this kind of relationship for poets and poetry, "agonistic"—and the lists seem pretty much the same for art in general. The imprecision of the term allows for my reciprocal methods, one dismissive (in note 29, of fancy alone for dimension i = 0) and the other inclusive (extending fancy to imagination for dimension i ≥ 0), as well as Bloom's revisionary ratios. But the extra resources in clinamen, etc. aren't necessary—the present-day rush of computer hype is merely an empty ploy (subterfuge) that's all smoke and mirrors, and only bluff and noise (endless shouting and vast commotion to attract attention). As 0-dimensional calculating, computers and AI/BIM are at best weak adversaries. Symbolic calculating is no match for visual calculating in shape grammars, and as a result, for insight and imagination in the embed-fuse cycle, and for art and design.

## EXHIBIT 2: OBSERVATIONS

1. This kind of dimensionality for shapes is a nice way to classify them and to distinguish shapes and symbols; it's discussed only briefly in my preface, and in "Seven Questions," in note 29. The canonical account is given in, G. Stiny, *Shape: Talking about Seeing and Doing* (Cambridge, MA: MIT Press, 2006), Parts I and II. The changes to calculating going from 0-dimensional things to 1-dimensions ones are simply remarkable.

2. G. Stiny, *Shape*, 296–301. The origin of this account is in G. Stiny, "Shape Rules: Closure, Continuity, and Emergence," *Environment and Planning B: Planning and Design* (1994): s49–s78. It's something of a surprise that continuity in shape grammars is defined retrospectively. For a little more incorporating identities x → x, see note 3 in "Seven Questions."

3. G. Stiny, *Pictorial and Formal Aspects of Shape and Shape Grammars* (Basel: Birkhäuser, 1975), 152–162. But see also G. Stiny, *Shape*, 186–188.

4. In each of my three examples and in fact, everywhere in this book, ambiguity spells the end of fixed intentions—there's always more to everything than anyone intends.

5. B. C. Smith, *On the Origin of Objects* (Cambridge, MA: MIT Press, 1996), 23–25. One of my students showed me Smith's example with two squares, to prove once and for all that shape grammars were fatally flawed. He was stunned when I embraced the squares and the accompanying rectangle effortlessly—shape grammars weren't what he thought they were, and his confusion was far from unique. It's funny how things aren't always what they seem to be at first blush. It was a "teachable moment"—where surprising opportunities arise to try something new, in the same way they do for shapes and rules in the embed-fuse cycle. But who says teaching is easy? My student knew what he knew, and wasn't about to change his mind—so much for shape grammars and the untold surprises they hold for art and design.

6. Not everything is explicated in advance in shape grammars either—but there's no need whatsoever for any kind of extra debugging every time something unexpected comes up or goes wrong. And what's not known now can be explained in retrospect, as a corollary of what rules see and do—without any kind of patch.

7. I was aware of gaps and spaces ("incomplete designs") early on, and the kinds of problems they were apt to cause, in G. Stiny, "Kindergarten Grammars: Designing with Froebel's Building Gifts," *Environment and Planning B* 7 (1980): 409–462, 443–445. I used a painfully obvious and shamefully tedious kind of debugging as a way around this, and viewed such side effects as a source of "unpleasantness" that might prove difficult to avoid. Looking back on what I was trying to do—to reconcile shapes, and parts that behaved like symbols—my impatience seems unjustified, and my methods seem pretty ad hoc, restricted and wrongheaded, something to deny and then to ignore immediately. Once invariant parts (as symbols) are picked out to define identities in the schema x → Σ F(prt(x)), unwelcome side effects disappear—it's no big deal to keep what you want. Of course, if you have any doubts about this—passing ones or not—it's probably a better bet to give up on what you think you want, in order to see where side effects go. Isn't the ability to assimilate surprise and delight a vital part of art and design, and why shapes supersede symbols, and seeing supersedes thought?

8. See, for example, G. Stiny, "A Note on the Description of Designs," *Environment and Planning B* 8 (1981): 257–267. This lays out the framework, but isn't as fluid as my current methods allow.

9. Some wonder if I have any favorite rules—which ones are full of delight? When I'm asked, I invariably point to the identities in the schema x → x. This is a neat way to highlight the distinction between object-oriented systems (computers) and shape grammars. On the one hand, identities aren't worth a second glance; they play no role when it comes to calculating with symbols, and in fact, they're better discarded, so as not to get in the way as they loop endlessly. Then on the other hand, identities strike at the quick of visual calculating, with embedding and shapes that fuse; at the very least, they pick out parts to alter what I see, and keep the parts/shapes I can't live without. Identities show how ambiguity works when I calculate with shapes and rules.

10. There are many arguments for this throughout G. Stiny, *Shape*—it's possible to start almost anywhere to find what you want. Owen Barfield distinguishes original and final participation in terms of figuration; they're both considered in A3 in "Seven Questions," 16.

11. G. Stiny, "*The Critic as Artist*: Oscar Wilde's Prolegomena to Shape Grammars," *Nexus Network Journal: Architecture and Mathematics* (2016): 1–36. I asked a good friend of mine in computer science to read this. After he finished, he asked me when I was first aware of the squares and triangles I use in Example 3, and that they weren't symbols. I replied that I knew about them when I was doing my PhD at UCLA in the theory of formal languages and automata, nearly 50 years ago. I didn't mention this at the time, because I thought it might be embarrassing. A lot of my work then was denying what I'd been taught and thought was true—and this is still pretty much the same now. Wilde would approve. My friend laughed—few today would care that there's an alternative way to calculate with shapes and rules that works in the embed-fuse cycle.

Symbolic calculating is perfectly fine the way it is, when everything is possible in computers and AI with data and machine learning. But somehow, this misses the point of pictures and poems (art and design) entirely, where Samuel Taylor Coleridge's esemplastic power (imagination) and Wilde's critical spirit hold sway. It's next to impossible to believe that prior data and learning can ever go beyond seeing things as in themselves they really are—motionless, hopelessly snared in numbers and statistics, objects essentially fixed and dead. After all, that's what's crucial when computers and AI take over the useful functions in our everyday lives, from running our cities to cleaning our houses to driving our cars, etc. There's little room for imagination and Wilde's critical spirit when you're crossing a busy intersection, at the risk of life and limb. Computers and AI can do many useful things within their broad compass, but only beyond these bounds are things vital and alive—in the region of John von Neumann's Rorschach test, in the region of beautiful form in art and design. Of course, there's nothing to keep computers from making pictures and poems, and no doubt, computer art is already a known reality—it's just not for computers and AI that can only see things as in themselves they really are, solely in terms of visual analogies and corresponding structures in graphs, trees, topologies, etc. These are defined from past experience with no recourse to new perception—mere data and learning seek what's invariant and leave out surprise. That's how computers and AI work, and what computer scientists prize the most—as scientists. There's a vast difference in sweep between symbolic calculating in computers and visual calculating in shape grammars—can anyone doubt the value of the latter in art and design? Without shape grammars, art and design are beyond what calculating can do. Maybe "what is art?"—computer art and not—isn't the right question; maybe "when is art?" makes better sense. It's art whenever someone looks at it and sees something surprising that he/she shouldn't, something strange to arouse aesthetic imagination. And this goes for artist and critic alike, making and judging—time after time.

12. J. E. Hopcroft, R. Motwani, and J. D. Ullman, *Introduction to Automata Theory, Languages, and Computation*, 3rd ed. (Boston: Pearson Education Inc., 2001), 256–259.

13. See the parentheses on pp. 45–47 of "Seven Questions" for the ins and outs of God, cybernetics, and shape grammars—all in a single song.

14. G. Stiny, *Shape*, 265–268. This is a nice way to see how finite groups work in terms of La Grange's theorem for subgroups and their cosets.

15. The popular distinction between thinking fast and thinking slow is key in Daniel Kahneman's and Amos Tversky's psychology, and in behavioral economics—D. Kahneman, *Thinking Fast and Slow* (New York: Farrar, Straus and Giroux, 2011). Thinking fast feels effortless—it's impulsive and intuitive, deciding things on the fly without knowing why. Thinking slow takes work—it's rational, sticking to prior facts with statistics and symbols to reason things out in logic and numbers, in order to calculate. On the face of it, seeing (perception) feels more like the former than the latter. In visual calculating, thinking fast goes best/always before thinking slow—what thinking slow (reasoning) offers relies first and foremost on thinking fast (seeing), to set the stage for description and representation. But thinking fast is often biased; it suppresses ambiguity to favor coherent stories from past experience that are stored in memory, saved for automatic use time and again—most of us share a polygonal bias that a triangle has three sides in a visual

analogy and that a square has four, etc. Left to itself, ambiguity is noise—I can see the same thing in many ways. It seems that ambiguities are recognized and resolved only thinking slow. Sometimes the practiced eye is frightfully slow, neither impulsive nor rational, taking its time to decide on the fly—thinking as in itself it really is not. I guess the right distinction isn't fast and slow. Of course, thinking fast and then slow with visual analogies have yet to show their true worth for my three Pythagorean rules, as they play out fully. There are many details to unfold, even in this simple example. The results are aptly summarized in an exotic immunological analogue—in fact, the immune system has been suggested as a practicable model for psychology in terms of nervous activity in the brain. Whether impulsive or rational, or neither one, the use of visual analogies is a lot like getting a flu shot; it usually works fine for the entire season, unless you're one of the unlucky ones who catches the flu. The vaccine can fail; it's never more than an educated (statistical) guess that may aim for the wrong strains. What actually happens skirts intentions and expectations, both expert ones and not. (Flu and vaccine, and shapes and symbols define another analogy to add to those in my preface—

flu : vaccine :: shapes : symbols

One of the remarkable strengths of current AI and machine learning is that they provide a new rule of three to solve verbal analogies, when words are represented in "thought vectors"—also see note 57 in "Seven Questions" for more fun. For example, in the partial analogy

Paris : _____ :: France : Italy

(Paris – France) + Italy = Rome. Neat, Paris is Rome—each is a European capital. But do they look and feel the same? I wonder how far this goes to get (flu – shapes) + symbols = vaccine, or for the original template

imagination : fancy :: shapes : symbols

in (imagination – shapes) + symbols = fancy, and its many surprising variations—what about (sin – shapes) + symbols = righteousness? But maybe the rule of three isn't meant to apply categorically; maybe there are unknown limits for thought vectors; maybe these are what von Neumann has in mind for visual analogies and the Rorschach test; maybe this is where computers, and AI and machine learning fail. Imagination poses similar problems anytime calculating ignores the embed-fuse cycle in shape grammars. This curbs art and design, plainly for analogy's kin—for figuration in pictures that replaces familiar visual analogies with new ones, and for metaphor in poems that swamps mappings and whatever relationships they may entail.)

16. G. Stiny, "Spatial Relations and Grammars," *Environment and Planning B* 9 (1982): 113–114. It's a real mystery to me why I preferred shapes to sets and seeing to counting—no one else did in the 1970s and 1980s, given the bent for scientific method and technical rigor in architecture and design research. Graphs and like structures were all the rage and in fact, still hold sway. (For a really nice review of the spirit of the 1970s and 1980s, at least in the Martin Centre at Cambridge, and the Open University, see note 16 in "Seven Questions.") I continue to be surprised by this today—that I'd much rather look at things and change my mind about them freely than say what they are, either in terms of visual analogies and spatial relations, or in terms of prior data and machine learning that make explicit visual analogies unnecessary. There's a lot more to seeing

than words, including graphs, etc. This can be a problem, especially when you're expected to talk way before you've had a chance to gawk.

17. Some straightforward examples of visual calculating in shape grammars, and combinatorial evidence that they can't be described/represented once and for all in terms of visual analogies (spatial relations or other kinds of computer structures in graphs, etc.), are given in I. Jowers, C. Earl, and G. Stiny, "Shapes, Structures, and Shape Grammar Implementation," *Computer-Aided Design* (2019): 80–92. The examples exploit the symmetry of squares, and their parallel sides. The rule

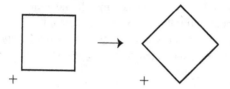

that rotates squares by π/4 around their centers is defined in the primary schema for transformations x → t(x). When this rule is applied to the shape

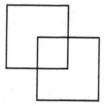

the sides of the three squares are structured in palindromic "words" or strings of incommensurable symbols/units. For b = a√2, the sides of the two large squares are segmented in the improbable string abababba. This also divides the sides of the small square in the palindrome ababa, with the palindromic prefix/suffix aba. But then, it's necessary to represent my rule twice—once for the large squares, and again for the small one. The prefix/suffix is crucial when the large squares are rotated

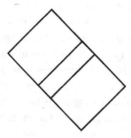

so that their sides align and overlap. In the common piece, the sides match each to the other, head to tail—hence, aba. The recursive structure-in-structure for palindromes in abababaaba at 1 (the large square), 1/2 (the small square), and $(2 - \sqrt{2})/2$ (the overlapping piece), however, is elusive in comparable shapes. In fact, such shapes are easy to define in indefinitely many ways, for example, when I add another small square in order to scale, either going up (putting in two larger squares) or equivalently going down (putting in one smaller square)

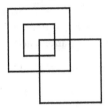

and when I push the large squares closer together along their common diagonal

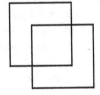

even by the tiniest nano-bit, but before the three squares merge/fuse into one. (As a pair of congruent squares, the second shape is part of a series that maps onto John McCarthy's lines—cantilevers and crosses—and 100 years before, Franz Reuleaux's nuts and bolts, where visual analogies vary continuously within fixed limits. McCarthy and Reuleaux are paradigmatic for parametric design.) Calculating runs smoothly without a hitch in the embed-fuse cycle with the rotation rule for squares and my pair of new shapes; it's impossible, however, with symbols that divide and segment from the start. This is probably no big surprise given how shape grammars work, although symbols are always ready for one more try and the promise of success the next time. To see segmenting loop endlessly can be mesmerizing, as it plays out in ever smaller and smaller divisions, and repeating patterns. There's no arguing with these kinds of combinatorial results, no matter that some of them spell misfortune and collapse for calculating in the usual way in computers—conspicuously in the object-oriented systems used/required in BIM and CAD. Of course, similar kinds of problems in object-oriented computer systems have already been considered in Example 2, in particular, for a square and a copy of it that form an undefined rectangle with sides that are 2 × 1. For this kind of problem in all shapes, there's a positive solution when reduction rules are used in the embed-fuse cycle. There's nothing to worry about in shape grammars—what you see is what you get. And in fact, that's the whole point of visual calculating, to be able to act on what you see, intended/defined and not.

18.  O. Wilde, "The Critic as Artist," *Intentions*, in *The Artist as Critic: The Critical Writings of Oscar Wilde*, ed. R. Ellmann (Chicago: University of Chicago Press, 1982), 349.

19.  G. Stiny, *Pictorial and Formal Aspects of Shape and Shape Grammars*, 226–228. See in particular the equilateral triangle in three squares and the square in four equilateral triangles drawn in Figure 2-35. (This is an immediate corollary of Proposition 2-1 for the inclusive set of shapes $\{-\}^{+}$, on p. 126—let a maximal line or any segment of one in any given shape be the aboriginal line in $\{-\}^{+}$. This substitution is straightforward using the identity for lines; it can be applied anywhere to keep the shape the same. Today, I'd go on to define the universe of shapes $\{-\}^{*}$ in a more designerly way that bypasses the line in $\{-\}^{+}$. I'd start out with the empty shape on a blank sheet of paper, and then apply the somewhat attenuated rule

that has the empty shape in its left-hand side, and a single maximal line is in its right-hand side. The rule renders every shape in a line drawing, pencil-stroke by pencil-stroke in a creative process that's independent of visual analogies, or any kind underlying structure or plan—it's a straight-forward trick in shape grammars to conjure something from nothing, simply by calculating. Now, $\{-\}^{+}$ is defined in reverse, when the empty shape is taken away from $\{-\}^{*}$.) The mystery of shapes isn't that there's so much in them, but that we see so little of it in ordinary looking—and are more than happy to settle for this, as soon as we can say anything about it. Words for visual analogies make a big difference. Three squares are three squares; they're named and defined—12 lines, four sides each—and that's all there is to it, even if my Pythagorean example and the A's just to come show the opposite. The part/role of artists (painters and poets) and designers is to see extravagantly—for rules in the embed-fuse cycle in shape grammars, this means looking beyond axioms and data, beyond what's given a priori, taught, or lodged in prior experience. That's where imagination kicks in, and where fancy can follow only after the fact.

20.  For an amusing discussion of this and how easy it is, even for the experienced thinker, to miss what's going on in verbal explication, instead of visual examples, see G. Stiny, *Shape*, 90–98. It's rollicking good fun to graph the equation $A + E = AE$, that is to say, $E = A/(A - 1)$, once you've seen that $A + E$ is the same as $AE$. It's kind of neat that E and A are undefined for $A = 1$ and $E = 1$. Then, 0 and 1 are as in themselves they really are not, each equals the other—try the math; it's not hard to check.

21.  J. D. Steingruber, *Architectural Alphabet 1773*, trans. E. M. Hatt (New York: George Braziller, Inc., 1975). It's amazing how easy it is to capture the observant eye—why not letters (symbols) for Leon Battista Alberti's tree-trunks and clods of earth? Charles Moore's notorious habit of collecting indigenous crafts and kitschy trinkets in his extensive travels is something of a marvel. But somehow, these artifacts invariably appear in his designs—additional tree-trunks and clods

of earth to see and copy, etc. In fact, Glenn Murcutt draws pretty much randomly to see what designs he can find in his scribbles. Less anecdotally, and with a nod more to painting and drawing than architecture, there's Leonardo da Vinci's stained walls and three centuries later, Alexander Cozens's "blots" that turn into landscapes, etc. in a visual process. (J. M. W. Turner's "Slave Ship" is an extension of this, as the eye scans an ominous sea of emergent/latent African faces, evoked by their haunting lips in the inverse schema for parts, that is to say $x \rightarrow prt^{-1}(x)$—at least that's how Turner's picture appears to me, so long as nothing takes away the looking. This mode of reverie overflows with meaning; the mood is simply overwhelming.) Such tricks aren't the exception, rather they're just how art and design work. Imagination/creativity is seeing, for the artist and the critic—in the embed-fuse cycle in shape grammars, with schemas and rules. Alberti is perfectly right about the unmatched power of the eye—to suppose that a bush is a bear, and all sorts of extravagant things in what there is to see.

22. Somehow, Steingruber's A in a series of parametric variations recalls the Vitruvian Man with arms and legs in motion. I used descriptions like this in a shape grammar for the *char-bagh*, or fourfold Mughul garden—G. Stiny and W. J. Mitchell, "The Grammar of Paradise: on the Generation of Mughul Gardens," *Environment and Planning B* 7 (1980): 209–226, 222. Today, I would recast the grammar entirely in terms of the schema $x \rightarrow \Sigma F(prt(x))$, for greater simplicity and improved clarity. The use of $x \rightarrow \Sigma F(prt(x))$ makes the grammar vastly easier than anything I initially thought of—all of the complication disappears in a kind of synoptic seeing and doing. I'm sure it's obvious by now, but I feel compelled to say it anyway—to be explicit, parametric variations like the ones I've defined for Steingruber's A and in Mughul gardens are neatly handled as transformations in schemas. Shape grammars subsume parametric design fully in every instance, and there's no reciprocity. In parametric design, shapes are 0-dimensional. This limits what rules can see and do in the embed-fuse cycle—only combinations of given units defined before calculating starts are ever possible. The set grammars in Example 3 and in note 16 in this exhibit show what's allowed.

23. I enjoy reading/rereading/misreading—they're all the same—the literary critic Harold Bloom, who tracks the influence of poet on poet in his fantastic "revisionary ratios." He notes in the preface of his last book, out a little while before his death, H. Bloom, *Possessed by Memory: The Inward Light of Criticism* (New York: Alfred Knopf, 2019), that he's lost in reverie and past argument, and that this is where he stands on the verge of his 10th decade, just beyond the threshold of his 90th year. (Bloom never tired of recounting his extravagant age.) Of course for the artist and the critic alike, the inevitable swerve from logical argument (visual analogy) to irony (re-description and open-ended representation) in unbounded reverie (re-constructed memory and evocation) is *de rigueur*—so it's odd and something of a surprise that Bloom feels compelled to affirm this in his writing, or at least to recognize it. Maybe there's the need to repudiate argument from axioms and data, in order to open up teaching in great universities like Bloom's Yale. For Wilde and certainly in shape grammars, the use of irony and reverie is perfectly normal, simply business as usual, to trace the locus of pictures and poems. In shape grammars, wherever this goes is part and parcel of shapes and rules, seeing and doing time after time in the embed-fuse cycle.

24. Personal e-mail correspondence with J. Gips (July 18, 2015), after a prerequisite lunch.

25. For a little more on my longtime aversion to hats that invariably rub and pinch, and feel like they're on even when they're off, see "Seven Questions," pp. 11–12.

### EXHIBIT 3: PEDAGOGY

1. The studio—writing about it can be an exercise in futility. Everyone has a different opinion about what's involved and why it matters, and maybe that's the point, to show that the studio is perfectly inclusive. Many tell me that I give the studio too much credit in art and design education—an architect learns on the fly, engaged in real projects. Others, in greater numbers, tell me that I'm not comprehensive enough—I focus on the aesthetics of art and design more than anything else, because that's where calculating seems least likely to make a meaningful difference, and to be honest, it's what interests me most. And this pleases no one, neither engineers nor aesthetes. In fact, aesthetes find the shameless intrusion of calculating super galling—maybe they'll get over it someday. But this isn't an excuse to give up. On the one hand, I'm not trying to analyze the studio, only to add to it without restricting it in any way—the general ways of the studio merit respect, secure enough for a second look in terms of visual calculating. And on the other hand, there seems to be a common feeling running through dissimilar accounts of the studio that it's an OK place to practice (try out) art and design, to participate in a creative community that's a democracy of the eye. That's the point of shape grammars, to let everyone calculate in terms of what they see, whatever it is—to be as inclusive as the studio at its best. That's why shapes are ambiguous, and the reason for the embed-fuse cycle—to guarantee inclusiveness, and full and open participation. My aim in this exhibit is to show that the eye fills the studio with content—that when it's tried in the embed-fuse cycle, visual calculating actually works to contribute everything the eye sees. My hope is that schemas for rules make up for any lapses in my account of the studio, especially for lapses that aren't in yours. Schemas add to the studio in spite of anything I say, whether positive, negative, or somewhere in between. That suggests another way to look at rules and how they work in the embed-fuse cycle, that they're indifferent to anything I've seen before—without ever being average.

2. For a delightful romp through some of the mumbo jumbo, see *Draw It with Your Eyes Closed: The Art of the Art Assignment* (Brooklyn, NY: Paper Monument, n+1 Foundation, Inc., 2012).

3. In the ferment of the studio, fancy is easy to confuse with fantasy and the fanciful—fancy may be fantasy or fanciful, or even fantastic and phantasmagoric, but it needn't be any of these. Fancy is combinatorial design with counters (Herbert Simon's "combinatoric play"); it leaves out imagination and lacks esemplastic power. When counters combine, they're independent and invariant, and never fuse and re-divide—what would counting be like if they did? Two centuries ago, Samuel Taylor Coleridge defined fancy in the *Biographia Literaria* (1817), and today this still goes for everything that relies on counters, things like computers with 0's and 1's—for machine learning and AI, for parametric design and BIM, and for divide and conquer, generate and test, heuristic search in humongous design spaces, and whatever's popular now. Also in the 19th century, Frederick Froebel invented the "kindergarten method" in sync with a vocabulary of children's play blocks presented in gifts (different counters to combine by eye and hand, at a personal

scale and pace)—in a productive rendition of fancy in education. In fact, the format of Froebel's kindergarten accounts for the studio in every guise—in general congruence and at times, in partial contrast. On the one hand, the studio mirrors the kindergarten in a special kind of free play that's influenced by a nurturing parental figure—a knowing studio master (he or she), a favorite teacher, or maybe even a psychiatrist. And then on the other hand, the eye veers off to flirt with the sublime in a rush of creative experience à la Coleridge, to see past vocabulary and fixed divisions in arrangements and groupings (sets) of separate counters, at least in the embed-fuse cycle that incorporates imagination and esemplastic power in calculating. The relationships coadunate in the kindergarten and the studio are resolved mutually in what follows in the main text.

4. William Blake's tiger and Charles Darwin's earthworm come up in "Seven Questions," note 50, in the contrast between romantic (aesthetic) enthusiasm and scientific (utilitarian) explanation—life in the former is independent of evolution in the latter. Darwin's theory is invoked in design in terms of genetic algorithms and such for optimization, usually in engineering and sometimes in architecture. This is a special case of shape grammars, in fact, a rather narrow one that crimps everything I'm trying for, in the studio and in art and design. But really, Darwin's locus in the embed-fuse cycle has yet to be traced—beyond an afterthought, when things are already fixed and dead, so that nothing evolves. More on this and how it might go is in note 9 below.

5. G. Stiny, "Kindergarten Grammars: Designing with Froebel's Building Gifts," *Environment and Planning B* 7 (1980): 409–462. Terry Knight extends this in her study of stylistic change—T. W. Knight, *Transformations in Design* (Cambridge: Cambridge University Press, 1994).

6. This involves composite shapes in products in G. Stiny, *Pictorial and Formal Aspects of Shapes and Shape Grammars* (Basel: Birkhäuser, 1975). And more explicitly, there's G. Stiny, "A Note on the Description of Designs," *Environment and Planning B: Planning and Design* 8 (1981): 257–267; G. Stiny, "What Is a Design," *Environment and Planning B: Planning and Design* 17 (1990): 97–103; and G. Stiny, "Weights," *Environment and Planning B: Planning and Design* 19 (1992): 413–430.

7. How visual calculating in shape grammars contrasts with "combinatoric play" in permutations and combinations is emphasized throughout "Seven Questions," and earlier in G. Stiny, *Shape: Talking about Seeing and Doing* (Cambridge, MA: MIT Press, 2006). Following the publication of "Kindergarten Grammars," I discussed this briefly in G. Stiny, "Spatial Relations and Grammars," *Environment and Planning B* 9 (1982): 113–114. But the stage was set in "Kindergarten Grammars," in the pair of schemas I used that includes every rule, 431, and in the three examples in figure 29, 436–437—in fact, everywhere in the kindergarten. There's no getting around the embed-fuse cycle in shape grammars. That's the reason they're indispensable for seeing and doing, and for art and design. Sometimes I wonder why no one tries the embed-fuse cycle in computers but then, who would ever suppose that imagination is something to calculate? No doubt, it's better to stick to the facts in spatial relations—John von Neumann's visual analogies are the same, with the same limitations.

8. For an early account of this and a foretaste of "Kindergarten Grammars," see G. Stiny, "Two Exercises in Formal Composition," *Environment and Planning B* 3 (1976): 187–210.

9. Combinatorial and mechanical relationships require strict adherence to the original counters and divisions in fancy—the originalist (intentional) imperative is wantonly ignored in imagination, in the embed-fuse cycle. There's more on this at the beginning of the preface in words and illustrations, and in the section near its end on Leon Battista Alberti and his famous motto QVID TVM. The folly of originalist strictures/structures in visual calculating is evident in the third example in Exhibit 2, in the contrast between invariant symbols in rules, and what happens with shapes. Things may not look the same now in the way they did originally; faithfulness to the past is no recipe for art and design, or for success elsewhere—in fact, it may be impossible to achieve. I use von Neumann to make a similar point in "Seven Questions" when he surveys the limits of calculating in pictures and the Rorschach test. And of course, Oscar Wilde shows how all of this goes together in a beautiful form—present meaning reconfigures original/past intentions, and may skirt them entirely. Imagination has no qualms.

10. What I have in mind is retrospective. How can you define a vocabulary of shapes/parts in advance unless you already know what you're going to do—is this really design? Doesn't it take away the looking? G. Stiny, "Shape Rules: Closure, Continuity, and Emergence," *Environment and Planning B: Planning and Design* 21 (1994): s49–s78; G. Stiny, "Useless Rules," *Environment and Planning B: Planning and Design* 23 (1996): 235–237. There's more on how to define a vocabulary of shapes in the following discussion of identities and how they're used to see, and trees and how to put them together in composite tables and graphs. Tellingly, vocabularies are a byproduct of descriptions that are easy to define with description rules (see above in note 6).

11. My hat's off to Ralph Waldo Emerson for "The eye is the best of artists"—not quite his "transparent eyeball," but super nonetheless—and lots more in "Nature," for example, in the transfiguring passion of the poet. That visual "delight" is "pleasure arising from outline, color, motion, and grouping," however, is pretty thin, and it's disappointing today given Coleridge's imagination and esemplastic power, and their full-bodied rendition in terms of schemas and rules for visual calculating in the embed-fuse cycle. To borrow/steal from another poet, the artist's eye follows rules in the embed-fuse cycle for the "silken skilled transmemberment" of everything it sees. This is the way to luxuriant delight. R. W. Emerson, "Nature," in *The Essential Writings of Ralph Waldo Emerson*, ed. B. Atkinson (New York: The Modern Library, 2000), 6, 8–9, 27.

12. There's a lot on schemas and many nice examples in the third part of *Shape*. Some of my expansive musings on how to define schemas are in G. Stiny, "What Rule(s) Should I Use?," *Nexus Network Journal* 13 (2011): 15–47. I'm easier about schemas today than I was when I started to play around with them, willing to accept any method that defines a set of rules in the embed-fuse cycle. The algebraic manipulation of schemas in terms of primary ones is appealing but without care, it turns quickly into a kind of symbol porn in a terrible addiction that's hard to break. The heuristic/intuitive and mnemonic/suggestive force of schemas is what counts most in the studio. Schemas, whether they're defined in formal expressions or with words and examples, are used for the same reasons to inform teaching and practice. Once again, the embed-fuse cycle is the real test. Schemas in words and examples are where Paul Klee comes in—in his *Pedagogical Sketchbook*, ed. S. Moholy-Nagy (New York: Frederick A. Praeger, 1953). Keeping to Klee's opening schemas, first there's "an active line, moving freely, without a goal, a walk for a walk's sake"

that's not unlike William Hogarth's specimen of the serpentine (see note 23 in "Seven Questions"), then "the same line, accompanied by [a] complementary form"

that tracks it's path with shortcuts, twists, and turns (similarly, a complementary form defines a building plan in two adjacent squares in the second example in Exhibit 2), and equally, "two secondary lines, moving around an imaginary main line"

in fact, the rudiments of an underlying grid to take away when it's no longer needed. It's no big deal to express these schemas in terms of mine, including my original schema x → A + B for spatial relations with play blocks in Froebel's building gifts, and it's inverse. But Klee is just one among many to appeal to words, and drawings and such to show schemas. I discuss this for a few others in *Shape*, 54, and in "What Rule(s) Should I Use?," 43; of course, there's Alberti's tree-trunks and clods of earth, and then Johann Wolfgang von Goethe's *Urpflanze* with organs that vary in terms of transformations and permutations (to paraphrase his maxim, all the organs of plants are leaves in parametric profusion), Gottfried Semper's *Urhutte* to survey buildings, Louis Sullivan's system of architectural ornament with its "seed-germ," and E. H. Gombrich's synoptic art and illusion—"The relationship between schemas [in] archetypical [drawings], and [formulas] to change them to meet the ongoing demands of experience and circumstance seems to be a very common idea in art and design that works with shapes, and the rules I define in my schemas to calculate." Incorporating schemas in studio teaching, in whatever way they're defined, isn't too much of a stretch. I guess I favor a relaxed version of my algebraic method with primary schemas and drawings, because this is usually more productive than words and examples—some things are simply too confused to say. Either way, there are drawings to figure out and then to misconstrue—and that's really the point of the embed-fuse cycle for primary schemas and words, as well. I'm happy to try anything that works to define rules in art and design, for anyone individually or for many in collaboration—to make it easier to do what you see, and sometimes

to say what this is. What's first and foremost is that you can always go on to something new—
effortlessly, no matter where you are.

13. I used the schemas $x \rightarrow x$ and $x \rightarrow x + t(x)$ in this way 45 years ago, when I started out with
shape grammars as an entry into art and design—in "Two Exercises in Formal Composition."
The addition schema $x \rightarrow x + t(x)$ is explicit in the rules for the first exercise; identities $x \rightarrow x$
are implicit in the second—sometimes, it takes a while to make sense of what you're doing and
sometimes, this is a surprise. When I wrote "Two Exercises," I knew all about generative grammars (Turing machines)—concatenation is what they do, so addition rules with triangles were
an obvious next step and easy to write out. Identities for symbols $A \rightarrow A$ were only to discard—
useless rules that only looped. There was a lot more yet to see.

14. G. Stiny and J. Gips, "Shape Grammars and the Generative Specification of Painting and
Sculpture," in *Information Processing 71*, ed. C. V. Freiman (Amsterdam: North-Holland, 1972),
1460–1465. A "generative specification" includes a "shape specification" in addition to a material one, to mirror Alberti's stark separation of plan (shape) and material in architectural design.
The same division goes in Monroe Beardsley's *Aesthetics: Problems in Philosophy of Criticism*, in the
sculptural equivalence between bronze and cheese of like hue—"let us ignore the smell," although
who would dare to say it's never noticeable. (Maybe sculpture has a limited self-life, and then
goes stale—isn't smell an infallible way to tell that it's spoiled?) This and some alternatives are
in G. Stiny and J. Gips, *Algorithmic Aesthetics: Computer Models for Criticism and Design in the Arts*
(Berkeley, CA: University of California Press, 1978), 28. But really, shape grammars are indifferent
to Alberti's distinction and Beardsley's olfactory version; it may or may not apply, and it usually
doesn't for shapes and weights.

15. Petra's drawings in Quist's studio are taken from D. A. Schön, *The Reflective Practitioner: How
Professionals Think in Action* (New York: Basic Books, 1983), 153–155. This is another way to copy
creatively. Schön would never view Petra and her drawings in terms of schemas, and calculating
with the rules they define in the embed-fuse cycle; he told me so stubbornly, in his self-assured
way, time and again. Schön viewed everything in terms of John Dewey, and was sure I miscon-
strued his logic of inquiry, especially my gloss of the studio as an open-ended (ambiguous)
process with indeterminate (ambiguous) results—so much for the spirit of inquiry. But let's give
Dewey an honest try—in his logic, inquiry (experiment) is

> the controlled or directed transformation of an indeterminate situation into to one that is so determinate in
> its constituent distinctions and relations as to convert the elements of the original situation into a unified
> whole.

This sounds like a garbled version of the embed-fuse cycle in which rules (identities) define
trees and graphs; it's close to William James who does better in the way he handles sagacity and
embedding ("Seven Questions," p. 26), and I. A. Richards is implicated somehow in his total
meaning (p. 43). But no one dares to say for sure—sometimes Dewey is hard to understand, lost
in his own words in an "indeterminate situation." Maybe Schön is right that inquiry is missing
when schemas are used in the studio—maybe inquiry is blind—or for Dewey, maybe inquiry can
follow the eye to do what it sees. To Schön, all calculating, visual and not, is only calculating in
computers—at best, computers are nothing more than useful tools for experienced professionals,

to validate their special expertise and recognized (rote) ways of doing things. But this doesn't extend to visual calculating in shape grammars. It's a cinch to be sentimental (romantic) about art and design in some pretty silly ways; they're outside calculating when your eyes are closed, and you aren't prepared to see—in fact, schemas in the embed-fuse cycle are all it takes to see in the studio, maybe even for inquiry.

16. The different rules I define for use in the summation schema $x \rightarrow \Sigma\ F(prt(x))$ also apply sequentially, but for the same result, I need the identities $x \rightarrow x$, as well. The details for this to work are spelled out near the end of the second example in Exhibit 2.

17. See figure 16 and the accompanying discussion in "Kindergarten Grammars," 426–427.

# INDEX